水利英语
ENGLISH FOR WATER RESOURCES

主　编　韩孟奇
副主编　庞彦杰　刘全勇　余桂霞　郑　宏

外语教学与研究出版社
FOREIGN LANGUAGE TEACHING AND RESEARCH PRESS
北京 BEIJING

图书在版编目 (CIP) 数据

水利英语 / 韩孟奇主编. —— 北京：外语教学与研究出版社，2021.6
ISBN 978-7-5213-2759-5

Ⅰ. ①水… Ⅱ. ①韩… Ⅲ. ①水利工程－英语－高等学校－教材 Ⅳ. ①TV

中国版本图书馆 CIP 数据核字 (2021) 第 129866 号

出 版 人　徐建忠
责任编辑　杨芳莉
责任校对　王　乐
版式设计　付玉梅
封面设计　锋尚设计
出版发行　外语教学与研究出版社
社　　址　北京市西三环北路 19 号（100089）
网　　址　http://www.fltrp.com
印　　刷　廊坊十环印刷有限公司
开　　本　787×1092　1/16
印　　张　10.5
版　　次　2021 年 9 月第 1 版 2021 年 9 月第 1 次印刷
书　　号　ISBN 978-7-5213-2759-5
定　　价　39.90 元

购书咨询：(010) 88819926　电子邮箱：club@fltrp.com
外研书店：https://waiyants.tmall.com
凡印刷、装订质量问题，请联系我社印制部
联系电话：(010) 61207896　电子邮箱：zhijian@fltrp.com
凡侵权、盗版书籍线索，请联系我社法律事务部
举报电话：(010) 88817519　电子邮箱：banquan@fltrp.com
物料号：327590001

前　言

《大学英语教学指南（2020版）》明确指出，大学英语课程兼具工具性和人文性。大学英语的工具性也体现在专门用途英语上，学生可以通过学习与专业或未来工作有关的学术英语或职业英语，获得在学术或职业领域用英语进行交流的相关能力。我们结合行业需求和学校特色，根据学校人才培养规格和学生需要，为水利行业院校学生编写了这本具有行业特色的英语教材，旨在将特定的学科内容与语言教学目标相结合，着重解决学生在用英语学习学科知识的过程中遇到的语言问题，重点培养与专业相关的英语阅读与翻译能力。

《水利英语》共设计十个单元，每单元两篇课文，选自水利工程中的重点环节，内容涉及水循环、水资源、河流与湖泊、水库、防洪、灌溉与排水、大坝、水轮机、水力发电、水环境与水污染等。为了保证教学效果，本教材还为每个单元提供了相关视频材料和在线资源。《水利英语》内容新颖，语言地道，难度适中，可以纳入水利行业院校通用英语教学体系或作为水利专业学生提高专业英语能力的教材。

本教材注重学生语言基本功的培养。教材的练习设计注重学生语言输出能力的提升，练习内容丰富，涵盖阅读、翻译和口语练习等，针对学生的实际需要，做到有的放矢。

本教材由华北水利水电大学韩孟奇担任主编，庞彦杰、刘全勇、余桂霞、郑宏担任副主编，由党兰玲教授负责主审工作。

本教材是河南省高等教育教学改革研究与实践项目："讲好中国故事"视域下大学英语"金课"研究（2019SJGLX284）的阶段性成果。

由于编者水平有限，书中难免有疏漏和不足之处，恳请读者和同行批评指正，以便再版时修正。

编者
2021年8月

Contents

Unit 1	**Water Cycle**	**1**
	Text A: Hydrologic Cycle	2
	Text B: Water's Journey Through Time	9
Unit 2	**Water Resources**	**17**
	Text A: Facts on Water Resources	18
	Text B: Types of Water Resources	26
Unit 3	**Lakes and Rivers**	**33**
	Text A: Freshwater Lakes and Rivers	34
	Text B: Rivers, Streams, and Creeks	42
Unit 4	**Reservoirs**	**47**
	Text A: Types of Reservoirs	48
	Text B: Three Gorges Reservoir Region	56
Unit 5	**Flood Control**	**63**
	Text A: Flood Control and Disaster Management	64
	Text B: More Floods, More Effective Flood-fighting Technology	71
Unit 6	**Irrigation and Drainage**	**79**
	Text A: Modern Irrigation System Planning and Construction	80
	Text B: Modern Drainage System Construction	87
Unit 7	**Dams**	**95**
	Text A: Dams Are a Vital Part of the National Infrastructure	96
	Text B: Dams: What They Are and What They Do	104

Unit 8	**Hydraulic Turbines**	**111**
	Text A: Hydro Turbine	112
	Text B: Factors Affecting the Selection of Hydraulic Turbines	118
Unit 9	**Hydroelectric Power**	**125**
	Text A: Hydropower Today	126
	Text B: Hydroelectric Energy and Its Pros and Cons	133
Unit 10	**Water Pollution**	**139**
	Text A: Water Pollution—Everything You Need to Know	140
	Text B: What Are the Effects of Water Pollution?	147
Appendix: Glossary		**154**

Unit 1
Water Cycle

Lead-in

Water cycle, also called hydrologic cycle, involves the continuous circulation of water in the Earth-atmosphere system. Of the many processes involved in the water cycle, the most important are evaporation, transpiration, condensation, precipitation, and runoff. Although the total amount of water within the cycle remains essentially constant, its distribution among the various processes is continuously changing.

Text A

Hydrologic Cycle

1 Water has been constant in quantity and continuously in motion. Little has been added or lost over the years. The same water molecules have been transferred time and time again from the oceans and the land surface into the atmosphere by evaporation, dropped on the land as precipitation, and transferred back to the sea by rivers and groundwater. This endless circulation is known as the "hydrologic cycle".

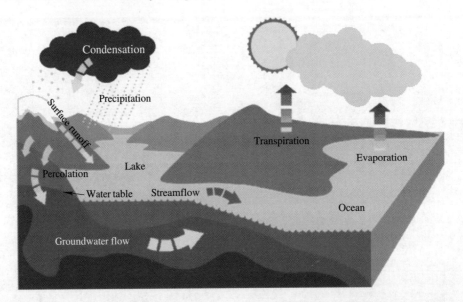

2 The illustration shows the hydrologic cycle in which water leaves the atmosphere and falls to Earth as precipitation where it enters surface waters or percolates into the water table and groundwater and eventually is taken back into the atmosphere by transpiration and evaporation to begin the cycle again.

Evaporation

3 As water is heated by the sun, surface molecules become sufficiently energized to break free of the attractive force binding them together, and then evaporate and rise as invisible vapor in the atmosphere.

Transpiration

4   Water vapor is also emitted from plant leaves by a process called transpiration. Every day an actively growing plant transpires 5 to 10 times as much water as it can hold at once.

Condensation

5   As water vapour rises, it cools and eventually condenses, usually on tiny particles of dust in the air. When it condenses it becomes a liquid again or turns directly into a solid (ice, hail or snow). These water particles then collect and form clouds.

Precipitation

6   Precipitation in the form of rain, snow and hail comes from clouds. Clouds move around the world, propelled by air currents. For instance, when they rise over mountain ranges, they cool, becoming so saturated with water that water begins to fall as rain, snow or hail, depending on the temperature of the surrounding air.

Runoff

7   Excessive rain or snowmelt can produce overland flow to creeks and ditches. Runoff is visible flow of water in rivers, creeks and lakes as the water stored in the basin drains out.

Percolation

8   Some of the precipitation and snowmelt moves downwards, percolates or infiltrates through cracks, joints and pores in soil and rocks until it reaches the water table where it becomes groundwater.

Groundwater

9   Subterranean water is held in cracks and pore spaces. Depending on the geology, the groundwater can flow to support streams. It can also be trapped by wells. Some groundwater is very old and may have been there for thousands of years.

Water table

10   The water table is the level at which water stands in a shallow well.

The sun-powered cycle

11 Heating of the ocean water by the sun is the key process that keeps the hydrologic cycle in motion. Water evaporates, then falls as precipitation in the form of rain, hail, snow, sleet, drizzle or fog. On its way to Earth some precipitation may evaporate or, when it falls over land, be intercepted by vegetation before reaching the ground. The cycle continues in three different ways:

- Evaporation/transpiration—On average, as much as 40 percent of precipitation in Canada is evaporated or transpired.

- Percolation into the ground—Water moves downward through cracks and pores in soil and rocks to the water table. Water can move back up by capillary action or it can move vertically or horizontally under the Earth's surface until it re-enters a surface water system.

- Surface runoff—Water runs overland into nearby streams and lakes; the steeper the land and the less porous the soil, the greater the runoff. Overland flow is particularly visible in urban areas. Rivers join each other and eventually form one major river that carries all of the sub-basins' runoff into the ocean.

12 Although the hydrologic cycle balances what goes up with what comes down, one phase of the cycle is "frozen" in the colder regions during the winter season. During the Canadian winter, for example, most of the precipitation is simply stored as snow or ice on the ground. Later, during the spring melt, huge quantities of water are released quickly, which results in heavy spring runoff and flooding.

The water-climate relationship

13 Water plays a basic role in the climate system through the hydrologic cycle, but water is intimately related to climate in other ways as well. It is obvious, from a water resource perspective, how the climate of a region to a large extent determines the water supply in that region based on the precipitation available and on the evaporation loss. Perhaps less obvious is the role of water in climate. Large water bodies, such as the oceans and the Great Lakes, have a moderating effect on the local climate because they act as a large source and sink for heat. Regions near these water bodies generally have milder winters and cooler summers than would be the case if the nearby water body did not exist.

14　The evaporation of water into the atmosphere requires an enormous amount of energy, which ultimately comes from the sun. The sun's heat is trapped in the Earth's atmosphere by greenhouse gases, the most plentiful of which by far is water vapor. When water vapor in the atmosphere condenses to precipitation, this energy is released into the atmosphere. Fresh water can mediate climate change to some degree because it is stored on the landscape as lakes, snow covers, glaciers, wetlands and rivers, and is a store of latent energy. Thus water acts as an energy transfer and storage medium for the climate system.

15　The water cycle is also a key process upon which other cycles operate. For example, one needs to properly understand the water cycle in order to address many of the chemical cycles in the atmosphere.

(919 words)

New Words

1	circulation /ˌsɜːkjəˈleɪʃən/	n.	循环
2	evaporation /ɪˌvæpəˈreɪʃən/	n.	蒸发
3	evaporate /ɪˈvæpəreɪt/	v.	（使）蒸发，挥发
4	propel /prəˈpel/	v.	推进
5	saturate /ˈsætʃəreɪt/	v.	使浸透，使充满，使饱和
6	runoff /ˈrʌnɒf/	n.	径流
7	ditch /dɪtʃ/	n.	沟渠，壕沟
8	creek /kriːk/	n.	小湾，小溪
9	percolate /ˈpɜːkəleɪt/	v.	渗入
10	infiltrate /ˈɪnfɪltreɪt/	v.	渗透
11	sleet /sliːt/	n.	雨夹雪
12	transpiration /ˌtrænspəˈreɪʃən/	n.	蒸发作用
13	transpire /trænˈspaɪə/	v.	蒸发
14	vapor /ˈveɪpə/	n.	水蒸气
15	condense /kənˈdens/	v.	冷凝
16	precipitation /prɪˌsɪpɪˈteɪʃən/	n.	降水
17	mediate /ˈmiːdieɪt/	v.	调和

Expressions

1	air current	气流
2	subterranean water	地下水
3	hydrologic cycle	水循环
4	water table	地下水位
5	capillary action	毛细管作用
6	sub-basin	子流域；分流域
7	moderating effect	调节效应
8	greenhouse gases	温室气体
9	latent energy	潜能
10	storage medium	存储介质
11	chemical cycle	化学循环

Notes

The Great Lakes 北美五大湖是世界最大的淡水湖群，即北美洲的苏必利尔湖、密歇根湖、休伦湖、伊利湖和安大略湖这五个相连湖泊的总称，又称大湖，有"北美大陆地中海"之称。北美五大湖除密歇根湖属于美国外，其余四湖均跨美国和加拿大两国。五大湖总面积超过24万平方千米，其中美国占72%，加拿大占28%。总蓄水容量约228,000亿立方米，约占全世界淡水湖总量的1/5。

Exercises

I. Read each of the following paraphrases and then write the word it represents on the line provided.

1. a small stream _____
2. a mixture of rain and snow or hail _____
3. something that drains or flows off, as rain water _____
4. to make more dense or compact _____
5. a substance that is in a gaseous state at a temperature below its boiling point _____
6. the process in which a liquid changes state to vapor _____
7. any form of water, such as rain, snow, sleet, or hail, that falls to the Earth's surface _____

8 (of plants) to lose (water in the form of water vapour), especially through the stomata of the leaves _____

9 to cause (a liquid, for example) to permeate a substance by passing through its pores _____

10 the gaseous envelope surrounding the Earth or a heavenly body; the air _____

II. Find the Chinese equivalents for the following expressions.
1 hydrologic cycle
2 air current
3 water table
4 capillary action
5 subterranean water
6 sub-basin
7 storage medium
8 chemical cycle
9 latent energy
10 moderating effect

III. Discuss the following questions.
1 How is water evaporated into the atmosphere?
2 What are the main forms of precipitation?
3 How is underwater formed according to the passage?
4 What are the different ways in which water cycle continues?
5 What role can large water bodies play in climate? Why?

IV. Choose the best answer to each of the following questions or statements.
1 In the hydrologic cycle water leaves the land surface into the atmosphere by _____.
   A. precipitation   B. evaporation
   C. air   D. water
2 What keeps the hydrologic cycle in motion?
   A. Wind.   B. Rain.
   C. Sun.   D. Water.

3　Plants can play a role in the hydrologic cycle by _____.
　　A. evaporation　　　　　　　　　B. percolation
　　C. condensation　　　　　　　　 D. transpiration
4　Which of the following is NOT the form of precipitation?
　　A. Hail.　　　　　　　　　　　　B. Sleet.
　　C. Hurricane.　　　　　　　　　 D. Snow.
5　Coastal regions or areas close to large water bodies tend to have _____ weather.
　　A. hot　　　　　　　　　　　　　B. cold
　　C. mild　　　　　　　　　　　　 D. rough

V. **Translate the following sentences into Chinese.**

1　Water vapor is also emitted from plant leaves by a process called transpiration.
2　As water vapor rises, it cools and eventually condenses, usually on tiny particles of dust in the air.
3　Heating of the ocean water by the sun is the key process that keeps the hydrologic cycle in motion.
4　Rivers join each other and eventually form one major river that carries all of the sub-basins' runoff into the ocean.
5　Water plays a basic role in the climate system through the hydrologic cycle, but water is intimately related to climate in other ways as well.

VI. **Translate the following sentences into English.**

1　水蒸气凝结时，会再变成液体或直接变成如冰雹、雪之类的固体。(condense)
2　显而易见，气候与水循环密切相关。(relate)
3　水蒸发进入大气需要很多能量，这些能量最终来自太阳。(evaporation)
4　降水的形式通常有下雨、下雪和冰雹。(precipitation)
5　沿海或大的湖泊附近地区的冬天不会太冷，夏天较凉爽。(mild)

VII. **Translate the passage into Chinese.**

It is obvious, from a water resource perspective, how the climate of a region to a large extent determines the water supply in that region based on the precipitation available and on the evaporation loss. Perhaps less obvious is the role of water in climate. Large water bodies, such as the oceans and the Great Lakes, have a moderating effect on the local climate because they act as a large source and sink for heat. Regions near these water bodies generally have milder winters and cooler summers than would be the case if the nearby water body did not exist.

Text B

Water's Journey Through Time

1. Earth is the only planet in our solar system with extensive liquid water—other planets are too hot or too cold, too big or too small. Though Mars appears to have had water on its surface in the past and may still harbor liquid water deep below its surface, our oceans, rivers, and rain are unique as far as we know, and they are life-sustaining. Understanding the processes and reservoirs of the hydrologic cycle is fundamental to dealing with many issues, including pollution and global climate change.

The Hydrologic Cycle

2. The hydrologic cycle can be thought of as a series of reservoirs, or storage areas, and a set of processes that cause water to move between those reservoirs. The largest reservoir by far is the oceans, which hold about 97% of Earth's water. The remaining 3% is the freshwater so important to our survival, but about 78% of that is stored in ice in Antarctica and Greenland. About 21% of freshwater on Earth is groundwater, stored in sediments and rocks below the surface of Earth. The freshwater that we see in rivers, streams, lakes, and rain is less than 1% of the freshwater on Earth and less than 0.1% of all the water on Earth.

The Ocean and the Atmosphere

3. Water moves constantly between these reservoirs through the processes of evaporation, condensation and precipitation, surface and underground flow, and others. The driving force for the hydrologic cycle is the sun, which provides the energy needed for evaporation just as the flame of a gas stove provides the energy necessary to boil water and create steam. Water changes from a liquid state to a gaseous state as it evaporates from the oceans, lakes, streams, and soil. Because the oceans are the largest reservoir of

liquid water, that is where most evaporation occurs. The amount of water vapor in the air varies widely over time and from place to place; we feel these variations as humidity.

4   The presence of water vapor in the atmosphere is one of the things that makes Earth livable for us. In 1859, Irish naturalist John Tyndall began studying the thermal properties of the gases in Earth's atmosphere. He found that some gases, like carbon dioxide (CO_2) and water vapor, trap heat in the atmosphere (a property commonly called the greenhouse effect), while other gases like nitrogen (N_2) and argon (Ar) allow heat to escape to space. The presence of water vapor in the atmosphere helps keep surface air temperatures on Earth in a range from about –40°C to 55°C. Temperatures on planets without water vapor in the atmosphere, like Mars, stay as low as –100°C.

5   Once water vapor is in the air, it circulates within the atmosphere. When an air package rises and cools, the water vapor condenses back to liquid water around particulates like dust, called condensation nuclei. Initially these condensation droplets are much smaller than raindrops and are not heavy enough to fall as precipitation. These tiny water droplets create clouds. As the droplets continue to circulate within the clouds, they collide and form larger droplets, which eventually become heavy enough to fall as rain, snow, or hail. Though the amount of precipitation varies widely over Earth's surface, evaporation and precipitation are globally balanced. In other words, if evaporation increases, precipitation also increases; rising global temperature is one factor that can cause a worldwide increase in evaporation from the world's oceans, leading to higher overall precipitation.

6   Since oceans cover around 70% of Earth's surface, most precipitation falls right back into the ocean and the cycle begins again. A portion of precipitation falls on land, however, and it takes one of several paths through the hydrologic cycle. Some water is taken up by soil and plants, some runs off into streams and lakes, some percolates into the groundwater reservoir, and some falls on glaciers and accumulates as glacial ice.

The Hydrologic Cycle on Land

7   The amount of precipitation that soaks into the soil depends on several factors: the amount and intensity of the precipitation, the prior condition of the soil, the slope of the landscape, and the presence of vegetation. These factors can interact in sometimes surprising ways—a very intense rainfall onto very dry soil, typical of the desert southwest,

often will not soak into the ground at all, creating flash-flood conditions. Water that does soak in becomes available to plants through soil moisture and groundwater. Plants take up water through their root systems, which mostly draw water from soil moisture; the water is then pulled up through all parts of the plant and evaporates from the surface of the leaves, a process called transpiration. Water that soaks into the soil can also continue to percolate down through the soil profile below the water table into groundwater reservoirs, called aquifers. Aquifers are often mistakenly visualized as great underground lakes; in reality, groundwater saturates the pore spaces within sediments or rocks.

8   Water that doesn't soak into the soil collects and moves across the surface as runoff, eventually flowing into streams and rivers to get back to the ocean. Precipitation that falls as snow in glacial regions takes a somewhat different journey through the water cycle, accumulating at the head of glaciers and causing them to flow slowly down valleys.

Humans and the Hydrologic Cycle

9   The properties of water and the hydrologic cycle are largely responsible for the circulation patterns we see in the atmosphere and the oceans on Earth. Atmospheric and oceanic circulation are two of the major factors that determine the distribution of climatic zones over the Earth. Changes in the cycle or circulation can result in major climatic shifts. For example, if average global temperatures continue to increase as they have in recent decades, water that is currently trapped as ice in the polar ice sheets will melt, causing a rise in sea level. Water also expands as it gets warmer, further exacerbating sea level rise. Many heavily populated coastal areas like New Orleans, Miami, and Bangladesh will be inundated by a mere 1.5 meter increase in sea level. Additionally, the acceleration of the hydrologic cycle (higher temperatures mean more evaporation and thus more precipitation) may result in more severe weather and extreme conditions. Some scientists believe that the increased frequency and severity of El Niño events in recent decades is due to the acceleration of the hydrologic cycle induced by global warming.

10 Even more immediately, the finitude of Earth's freshwater resources is becoming more and more apparent. Groundwater can take thousands or millions of years to recharge naturally, and we are using these resources far faster than they are being replenished. The water table in the Ogallala Aquifer, which underlies 175,000 square miles of the U.S. from Texas to South Dakota, is dropping at a rate of 10–60cm per year due to extraction

for irrigation. Surface waters around the world are largely contaminated by human and animal waste, most noticeably in countries like India, where untreated rivers provide the drinking and washing water for more than one billion people. Although legislation like the Clean Water Act in the United States and water conservation practices such as the use of low-flow toilets and showerheads in parts of the world has begun to address these issues, the problems will only grow as world population increases. Every spring and well, every river and sea does indeed flow from the same source, and changes affect not just one river or lake, but the whole hydrologic cycle.

(1,239 words)

New Words

1. harbor /ˈhɑːbə/ v. 包含
2. reservoir /ˈrezəvwɑː/ n. 水库
3. sediment /ˈsedɪmənt/ n. 沉积物
4. flame /fleɪm/ n. 火焰
5. humidity /hjuːˈmɪdɪti/ n. 湿度
6. nitrogen /ˈnaɪtrədʒən/ n. 氮
7. argon /ˈɑːgɒn/ n. 氩
8. particulate /pɑːˈtɪkjəleɪt/ n. 微粒状物质
9. accumulate /əˈkjuːmjʊleɪt/ v. 积累，积聚
10. aquifer /ˈækwɪfə/ n. 含水层
11. visualize /ˈvɪʒuəlaɪz/ v. 设想
12. inundate /ˈɪnəndeɪt/ v. 淹没
13. finitude /ˈfɪnɪtjuːd/ n. 有限性
14. recharge /ˌriːˈtʃɑːdʒ/ v. 恢复
15. replenish /rɪˈplenɪʃ/ v. 补充，再装满
16. underlie /ˌʌndəˈlaɪ/ v. 成为……的基础
17. extraction /ɪkˈstrækʃən/ n. 抽取
18. contaminate /kənˈtæmɪneɪt/ v. 污染
19. showerhead /ˈʃaʊəhed/ n. 莲蓬式喷头

Unit 1 Water Cycle

Expressions

1	condensation nuclei	凝结核
2	flash-flood	暴洪
3	soil profile	土壤剖面
4	circulation pattern	环流模式
5	climatic zone	气候带
6	climatic shift	气候变化
7	ice sheet	冰原，冰盖
8	global warming	全球变暖

Notes

1. **Greenland** 格陵兰岛，位于北美洲东北部，是丹麦属地之一，世界上最大的岛屿。格陵兰岛地处于北美洲与欧洲的交界处，沟通了北冰洋和大西洋，西部与加拿大隔海峡相望，北部濒临北冰洋，南部濒临大西洋，东部通过丹麦海峡与欧洲的冰岛隔海相望。
2. **The Clean Water Act** 清洁水法案。1972年，美国联邦政府颁布了旨在治理水污染的清洁水法案，该法案主要包括两大部分：第一部分是授权联邦政府对于市政污水处理设施建设提供资金援助；第二部分是关于工业和市政污水排放的法律规范和执法措施。清洁水法案大大减少了美国的水资源污染。

Exercises

I. Read each of the following paraphrases and then write the word it represents on the line provided.

   1 to house or contain _____
   2 a place where a great stock of anything is accumulated _____
   3 a measure of the amount of water vapor in the atmosphere _____
   4 to make impure or unclean by contact or mixture _____
   5 to charge again with electricity _____

II. Find the Chinese equivalents for the following expressions.

   1 condensation nuclei
   2 flash-flood
   3 circulation patterns

4　climatic zone

5　climatic shift

6　ice sheet

III.　**Discuss the following questions.**

1　Why will global warming have a great influence over the water cycle?

2　Why is water vapor in the atmosphere indispensable for the living of human beings?

3　Where does most precipitation happen? Why?

4　In water cycles, what usually happens to precipitation falling on land?

5　What will happen if the acceleration of the hydrologic cycle continues?

IV.　**Decide whether the following statements are true or false according to the text.**

_____ 1　Other planets besides Earth in the solar system are likely to have extensive liquid water.

_____ 2　As far as we know, the largest reservoir on Earth is the oceans.

_____ 3　Water vapor plays a critical role in producing greenhouse effect in the atmosphere.

_____ 4　Global temperature falling is likely to lead to higher overall precipitation.

_____ 5　Wind is the main driving force of the water cycle.

_____ 6　Plants on Earth can make some contributions to precipitation.

_____ 7　Global rising temperature may result in more severe weather around the world.

_____ 8　The increase of world population may lead to the decrease of global water amount.

_____ 9　The use of low-flow toilets and showerheads can help to solve the issue of fresh water resources to some extent.

_____ 10　The hydrologic cycle may be affected by human beings.

V.　**Translate the following sentences into English.**

1　就像煤气炉的火焰为烧开水提供所需的能量那样，太阳为水循环提供能量。(provide)

2　海洋是最大的液态水储存库，大部分水分蒸发发生在海洋。(occur)

3　全球气温上升会引起蒸发量的增加，从而导致更大的降水量。(global temperature)

4　植物通过根系吸收水分，而根系主要是从土壤中吸取水分。(take up)

5　如果全球平均气温继续上升，极地地区的冰雪将会融化，导致海平面上升。(melt)

VI. **Translate the passage into Chinese.**

Earth is the only planet in our solar system with extensive liquid water—other planets are too hot or too cold, too big or too small. Though Mars appears to have had water on its surface in the past and may still harbor liquid water deep below its surface, our oceans, rivers, and rain are unique as far as we know, and they are life-sustaining. Understanding the processes and reservoirs of the hydrologic cycle is fundamental to dealing with many issues, including pollution and global climate change.

Unit 2
Water Resources

Lead-in

Water resources are sources of water that are useful or potentially useful to humans. Water is important because it is needed for life to exist. Many uses of water include agricultural, industrial, household, recreational and environmental activities. Virtually all of these human uses require fresh water. Only 2.5% of water on the Earth is fresh water, and over two thirds of this is frozen in glaciers and polar ice caps. Water demand already exceeds supply in many parts of the world, and many more areas are expected to experience this imbalance in the near future.

Text A

Facts on Water Resources

1. Water is essential for human survival and well-being and important to many sectors of the economy. However, resources are irregularly distributed in space and time, and they are under pressure due to human activity. How can water resources be managed sustainably while meeting an ever-increasing demand?

2. Around the world, human activity and natural forces are reducing available water resources. Although public awareness of the need to better manage and protect water has grown over the last decade, economic criteria and political considerations still tend to drive water policy at all levels. Science and best practice are rarely given adequate consideration.

3. Pressures on water resources are increasing mainly as a result of human activity—namely urbanization, population growth, increased living standards, growing competition for water, and pollution. These are aggravated by climate change and variations in natural conditions.

4. Still, some progress is being made. More and more, officials are evaluating water quantity and quality together, and coordinating management efforts across borders.

5. The world's water exists naturally in different forms and locations: in the air, on the surface, below the ground, and in the oceans.

6. Freshwater accounts for only 2.5% of the Earth's water, and most of it is frozen in glaciers and icecaps. The remaining unfrozen freshwater is mainly found as groundwater, with only a small fraction present above ground or in the air.

7. Looking at how water moves through the Earth's water cycle helps us understand how it interacts with the environment and how much is available for human use.

8. Precipitation—rain, snow, dew etc.—plays the key role in renewing water resources and in defining local climatic conditions and biodiversity. Depending on the local conditions, precipitation may feed rivers and lakes, replenish groundwater, or return to the air by evaporation.

9   Glaciers store water as snow and ice, releasing varying amounts of water into local streams depending on the season. But many are shrinking as a result of climate change.

10 River basins are a useful "natural unit" for the management of water resources and many of them are shared by more than one country. The largest river basins include the Amazon and Congo Zaire basins. River flows can vary greatly from one season to the next and from one climatic region to another. Because lakes store large amounts of water, they can reduce seasonal differences in how much water flows in rivers and streams.

11 Wetlands—including swamps, bogs, marshes, and lagoons—cover 6% of the worlds land surface and play a key role in local ecosystems and water resources. Many of them have been destroyed, but the remaining wetlands can still play an important role in preventing floods and promoting river flows.

12 Of the freshwater which is not frozen, almost all is found below the surface as groundwater. Generally of high quality, groundwater is being withdrawn mostly to supply drinking water and support farming in dry climates. The resource is considered renewable as long as groundwater is not withdrawn faster than nature can replenish it, but in many dry regions the groundwater does not renew itself or only very slowly. Few countries measure the quality of groundwater or the rate at which it is being exploited. This makes it difficult to manage.

13 The quantity of freshwater that is available for use by a given country in a given year, without exceeding the rate at which it is renewed, can be estimated taking into account the amount of precipitation, water flows entering and leaving the country, and water shared with other countries.

14 The average amount available per person varies from less than 50 m³ per year in parts of the Middle East to over 100,000 m³ per year in humid and sparsely populated areas.

15 The United Nations has kept a country by country database of such estimates for several decades.

16 Though the database has become a common reference tool, it has some drawbacks. Figures only indicate the maximum theoretical amount available for a country and may be an overestimation. Moreover, annual and national averages tend to mask local and seasonal differences.

17 Water resources face a host of serious threats, all caused primarily by human activity. They include pollution, climate change, urban growth, and landscape changes such as deforestation. Each of them has its own specific impact, usually directly on ecosystems and in turn on water resources.

18 If inadequately managed, activities like farming, forest-clearing, road-building, and mining can lead to too much soil and suspended particles ending up in rivers (sedimentation). This damages aquatic ecosystems, impairs water quality and hinders inland shipping.

19 Pollution can harm water resources and aquatic ecosystems. Major pollutants include for instance organic matter and disease-causing organisms from waste water discharges, fertilizers and pesticides running off from agricultural lands, acid rain resulting from air pollution, and heavy metals released by mining and industrial activities.

20 The effects of extracting too much water, both from surface waters and groundwater, have been dramatic. A striking example is the drastic reduction in size of the Aral Sea and Lake Chad. Little is being done to address the causes, which include poor water management practices and deforestation. In recent decades, much more water has been extracted from underground sources. The benefits of withdrawing groundwater are often short-lived, while the negative consequences—lower water levels and depleted resources, for example—can last a long time.

21 Climate change appears to increase existing pressures, for example in areas already suffering from water shortages. Land and mountain glaciers have been shrinking more rapidly in recent years. Extreme weather events stemming from global warming, such as storms and floods, are likely to become more frequent and severe. However, based on current knowledge, scientists can only make general predictions about the impact of climate change on water resources.

22 Meeting a continuous and ever-increasing demand for water requires efforts to compensate for natural variability, and to improve the quality and quantity available.

23 Rainwater has been collected for thousands of years in many parts of the world. Today, this technique is used in Asia to replenish underground supplies. It is relatively inexpensive and has the advantage of allowing local communities to develop and maintain the required structures themselves.

24. Diverting surface water into the ground can help reduce losses from evaporation, compensate for variations in flow, and improve quality. Middle East and Mediterranean regions are applying this strategy.

25. Dams and reservoirs have been built to store water for irrigation and drinking. Moreover dams can provide power and help control floods, but they can also bring about undesirable social and environmental impacts.

26. Transferring water between river basins can also help alleviate shortages. China, for instance, already has major interbasin links, and is planning more. The impact of these projects on people and the environment must be monitored closely.

27. Wastewater is now reused for different purposes in many countries, especially in the Middle East, and this practice is expected to grow. Worldwide, non-potable water is used for irrigation and industrial cooling. Cities are also turning to water re-use to supplement drinking water supplies, taking advantage of progress in water treatment.

28. Desalinated water—seawater and other salty water that has been turned into freshwater—is used by cities and by industries, especially in the Middle East. The cost of this technique has dropped sharply, but it relies heavily on energy from fossil fuels and raises waste management and climate change issues.

29. Using water resources sustainably is challenging because of the many factors involved, including changes in climate, the natural variability of the resource, as well as pressures due to human activity.

30. At present, most water policy is still driven by short-term economic and political concerns that do not take into account science and good stewardship. State-of-the-art solutions and more funding, along with more data on water resources, are needed especially in developing nations.

31. To assess the state of our water resources, we must fully appreciate the roles of different parts of the water cycle—such as rain, meltwater from glaciers, and so on. Otherwise, it remains difficult to develop adequate protection and mitigation strategies.

32. Poor water quality and unsustainable use of water resources can limit the economic development of a country, harm health and affect livelihoods. More sustainable practices are starting to be adopted.

33 When managing water resources, more attention should be paid to increasing existing natural resources and reducing demand and losses. The traditional response to rising demand for water was to store surface water in reservoirs, divert flow to dry regions and withdraw groundwater. Now these methods are increasingly supplemented by water reuse, desalination and rainfall harvesting. Certain regions are even going to the extreme of exploiting non-renewable groundwater resources.

34 Some countries have programs to reduce demand and losses from urban water distribution systems but more efforts are necessary. However, this will involve changes in behaviors requiring education and political commitment. Such efforts to conserve water and reduce demand are not only useful in regions where water is in short supply, they can also bring economic benefits in wetter regions.

35 Decentralized approaches to water resource management that focus on river basins are increasingly pursued even across borders. Exchanging information between countries that share river basins will yield both economic and environmental benefits.

(1,519 words)

New Words

1. aggravate /ˈæɡrəveɪt/ v. 恶化
2. glacier /ˈɡlæsiə/ n. 冰川
3. swamp /swɒmp/ n. 沼泽
4. bog /bɒɡ/ n. 沼泽
5. marsh /mɑːʃ/ n. 沼泽
6. lagoon /ləˈɡuːn/ n. 小而浅的湖
7. sparsely /ˈspɑːsli/ adv. 稀疏地，贫乏地
8. drawback /ˈdrɔːbæk/ n. 缺点
9. deforestation /diːˌfɒrɪˈsteɪʃən/ n. 毁林，滥伐森林
10. sedimentation /ˌsedɪmenˈteɪʃən/ n. 沉积，沉降
11. aquatic /əˈkwætɪk/ adj. 水生的
12. impair /ɪmˈpeə/ v. 损害，削弱

13	fertilizer /ˈfɜːtɪlaɪzə/ n.	肥料
14	pesticide /ˈpestɪsaɪd/ n.	杀虫剂，农药
15	drastic /ˈdræstɪk/ adj.	激烈的
16	deplete /dɪˈpliːt/ v.	耗尽
17	compensate /ˈkɒmpənseɪt/ v.	补偿
18	alleviate /əˈliːvieɪt/ v.	减轻，使……缓和
19	desalinate /diːˈsælɪneɪt/ v.	淡化（海水），（从海水中）除去盐分
20	stewardship /ˈstjuːədʃɪp/ n.	管理工作
21	mitigation /ˌmɪtɪˈɡeɪʃən/ n.	缓和，减轻
22	divert /daɪˈvɜːt/ v.	转移

Expressions

1	river basin	流域；流域盆地
2	suspended particles	悬浮颗粒
3	aquatic ecosystem	水生生态系统
4	non-potable water	非饮用水
5	state-of-the-art	最新式的；使用了最先进技术的
6	rainfall harvesting	雨水收集

Notes

1. **Aral Sea** 咸海是位于中亚的咸水湖，湖水主要依赖阿姆河和锡尔河补充，但如今阿姆河基本不流进咸海了。
2. **Lake Chad** 乍得湖是非洲第四大湖，内陆淡水湖，位于非洲中北部，乍得盆地中央；由大陆局部凹陷而成，为第四纪古乍得海的残余。

Exercises

I. Read each of the following paraphrases and then write the word it represents on the line provided.

　　1　a mass of slowly moving land ice formed by the accumulation of snow on high ground

2 the cutting down and removal of all or most of the trees in a forested area _____

3 to make (pain, for example) less intense or more bearable _____

4 a substance or agent used to kill pests, such as unwanted or harmful insects, rodents, or weeds _____

5 to consume or reduce to a very low amount _____

II. **Find the Chinese equivalents for the following expressions.**

1 river basin

2 aquatic ecosystem

3 non-potable water

4 rainfall harvesting

5 suspended particles

III. **Discuss the following questions.**

1 Where and in what forms is water available on Earth?

2 How can we estimate the quantity of freshwater that is available for use by a given country in a given year?

3 In what ways can human actions affect water resources?

4 How can the growing need for water be met?

5 How could water resources be developed sustainably?

IV. **Choose the best answer to each of the following questions or statements.**

1 Groundwater is considered renewable as long as _____.

   A. nature can replenish groundwater slower than it is withdrawn

   B. nature can replenish groundwater faster than it is withdrawn

   C. the groundwater in many dry regions does not renew itself

   D. the groundwater in many dry regions renews itself very slowly

2 Which one does NOT belong to the drawbacks of the water resources database of the UN?

   A. Figures only indicate the maximum theoretical amount available for a country.

   B. Figures may be an overestimation.

   C. The database has become a common reference tool.

   D. Annual and national averages tend to mask local and seasonal differences.

3 Which one is the cause of the drastic reduction in size of the Aral Sea and Lake Chad?

   A. Fertilizers and pesticides running off from agricultural lands.

   B. Heavy metals released by mining and industrial activities.

C. Poor water management practices and deforestation.

D. Acid rain resulting from air pollution.

4 Desalinated water is used by cities and by industries, especially in _____.

A. China
B. Mediterranean regions
C. the Amazon
D. the Middle East

5 What is the focus of decentralized approaches to water resource management?

A. River basins.
B. Dams.
C. Urban water distribution systems.
D. Rainwater.

V. **Translate the following sentences into Chinese.**

1 Depending on the local conditions, precipitation may feed rivers and lakes, replenish groundwater, or return to the air by evaporation.

2 If inadequately managed, activities like farming, forest-clearing, road building, and mining can lead to too much soil and suspended particles ending up in rivers (sedimentation).

3 The benefits of withdrawing groundwater are often short-lived, while the negative consequences—lower water levels and depleted resources, for example—can last a long time.

4 Diverting surface water into the ground can help reduce losses from evaporation, compensate for variations in flow, and improve quality.

5 At present, most water policy is still driven by short-term economic and political concerns that do not take into account science and good stewardship.

VI. **Translate the following sentences into English.**

1 不同季节之间以及不同气候区域之间的河流流量差异很大。(vary)

2 水资源面临着一系列严重的威胁，这些威胁主要是由人类活动造成的。(face)

3 从地表水和地下水中抽取过多的水，其影响是巨大的。(extract from)

4 这项技术的成本大幅下降，但它引发了废物管理和气候变化问题。(raise)

5 要评估我们的水资源状况，我们必须充分认识到水循环不同部分的作用。(assess)

VII. **Translate the passage into Chinese.**

Of the freshwater which is not frozen, almost all is found below the surface as groundwater. Generally of high quality, groundwater is being withdrawn mostly to supply drinking water and support farming in dry climates. The resource is considered renewable as long as groundwater is not withdrawn faster than nature can replenish it, but in many dry regions the groundwater does not renew itself or only very slowly. Few countries measure the quality of groundwater or the rate at which it is being exploited. This makes it difficult to manage.

Text B

Types of Water Resources

1. Around 71 percent of the Earth's surface is covered in water. This massive quantity of water is hard to visualize: the total water resources of the Earth equal roughly 326 million cubic miles, with each cubic mile equal to around 1 trillion gallons of water. To imagine just one trillion gallons of water, try to picture 40 million swimming pools, or 24 billion baths. Now, multiply those numbers by 326 million!

2. Of all of this water, only about 2.5 percent is freshwater: the other 97.5 percent is saltwater. Almost 69 percent of freshwater resources are tied up in glaciers and ice caps, about 30 percent is groundwater, and a mere 0.27 percent is surface water. While all kinds of water resources are important for the survival of the planet, accessible freshwater is especially important for humans.

Saltwater Resources

3. As mentioned, saltwater is abundant in the surface of the planet. However, saltwater is currently not particularly useful when it comes to potable water supplies. Desalination plants, while they do exist, are scarce because the energy required for desalination makes the process extremely expensive.

4. That said, there are saltwater resources from which humans benefit, aside from beautiful ocean views. Saltwater fish are a staple in much of the world's diet (although overfishing and pollution has put much of the marine life population at risk). Furthermore, tidal waters are being used as a source of hydroelectric energy. So, while saltwater is not helpful in dealing with scarce water supplies, it does provide resources that humans rely on.

Groundwater Resources

5   Groundwater is the most plentiful of all freshwater resources. As water percolates into the ground through layers of soil, clay, and rock, some of it adheres to the topmost layers to provide water to plants. This water is in what is called the unsaturated, or vadose, zone. Most of the pores in the vadose zone are filled with air, rather than water.

6   Gravity continues to move the water down through the ground. Eventually, the water reaches the saturated zone, where all the pores are filled with water. The separation between the saturated and unsaturated zone is called the water table.

7   Aquifers are areas of permeable rock that hold water. Typically, aquifers are made of bedrock that has many fractures and connected pores, such as limestone, sandstone and gravel. Shale and clay layers are impermeable, and therefore make poor aquifers. An aquifer is "recharged" through precipitation from above percolating through the layers of soil and rock. Therefore, there is significant interaction between surface water and ground water.

8   In turn, groundwater feeds surface water through springs, and surface water can also recharge groundwater supply.

9   Most often, groundwater is accessed by humans via wells. To build a well, one must drill down past the water table. In most cases, a pump is placed in the bottom of the well, and it is pumped into homes, businesses and water treatment plants, where it is then dispersed. As water is pumped from the ground, a cone of depression forms around the well. The groundwater from the surrounding area moves towards the well. Wells can run dry during times of drought, or if surrounding wells are pumping too much water, causing the cone of depression to be large.

10 Water pumped from wells is generally very clean. The layers of soil, clay and rock acts as a natural filter. However, contaminants from nearby contaminated soils, leaky underground tanks, and septic systems can pollute a well, rendering it unusable. Furthermore, salt water intrusion can occur when the rate of pumping near a shoreline exceeds the rate of recharge. Saltwater gets pulled from the ocean into the cone of depression, and enters the well.

11 Subsidence, the gradual settling of the land due to continuous pumping and development, has also become an issue as groundwater is mined. This occurs when groundwater

is pumped out faster than it can be replenished, and the sediment beneath becomes compacted. Subsidence is a permanent phenomenon. It can cause structural problems to foundations, an increased incidence of sinkholes and flooding problems. To top it off, subsidence is extremely costly. In some areas, such as the San Joaquin Valley in California, the land has subsided over 30 feet due to groundwater withdraw.

Surface Water Resources

12 Surface water is the water that exists in streams and lakes. This water is primarily used for potable water supply, recreation, irrigation, industry, livestock, transportation and hydroelectric energy. Over 63 percent of the public water supply is withdrawn from surface water. Irrigation gets 58 percent of its water supply from surface water. Industry gets almost 98 percent of its water from surface water systems. Therefore, surface water conservation and quality are of the utmost importance.

13 Watershed organizations continuously measure the stream flow and quality of surface water. Stream flow is monitored to warn of flooding and drought conditions. Water quality is very important, as the majority of the water used in the United States comes from surface water. It is the measure of how suitable the water is from a biological, chemical and physical perspective. Water quality can be impacted negatively by both natural and human causes: electrical conductivity, pH, temperature, phosphorus levels, dissolved oxygen levels, nitrogen levels and bacteria are tested as a measure of water quality.

14 Water that runs off into the stream can naturally carry sediment, debris and pathogens. Turbidity, the measure of suspended sediment in a stream, is also a measure of water quality. The more turbid the water, the lower the water quality.

15 Man-made contaminants such as gasoline, solvents, pesticides, and nitrogen from livestock can wash over the land and can leach into waterways, degrading the quality of nearby waters. The Clean Water Act in the United States protects the quality of the stream and issues fines to those contributing to the degradation in water quality. By protecting and conserving the water supply, there is a greater guarantee of future water resources for human use.

(985 words)

New Words

1. potable /ˈpəutəbəl/ adj.   适于饮用的
2. staple /ˈsteɪpəl/ n.   主食
3. unsaturated /ʌnˈsætʃəreɪtɪd/ adj.   不饱和的
4. pore /pɔː/ n.   孔隙
5. vadose /ˈveɪdəus/ adj.   渗流的
6. permeable /ˈpɜːmiəbəl/ adj.   有渗透性的
7. bedrock /ˈbedrɒk/ n.   基岩（松软的沙、土层下的岩石）
8. shale /ʃeɪl/ n.   页岩，泥板岩
9. disperse /dɪˈspɜːs/ v.   分散，扩散
10. cone /kəun/ n.   圆锥体，锥形物
11. contaminant /kənˈtæmɪnənt/ n.   杂质，污染物
12. intrusion /ɪnˈtruːʒən/ n.   入侵，闯入
13. subsidence /səbˈsaɪdəns/ n.   沉降，下沉
14. sinkhole /ˈsɪŋkhəul/ n.   落水洞（尤指石灰石地面下陷，地表水消失于地下）
15. livestock /ˈlaɪvstɒk/ n. pl.   家畜
16. phosphorus /ˈfɒsfərəs/ n.   磷
17. debris /ˈdebriː, ˈdeɪ-/ n.   碎片，残骸
18. pathogen /ˈpæθədʒən/ n.   病原体
19. turbidity /tɜːˈbɪdɪti/ n.   混浊度
20. solvent /ˈsɒlvənt/ n.   溶剂

Expressions

1. surface water   地表水
2. clay layer   黏土层
3. ground water   地下水
4. septic system   污水处理系统

> **Notes**
> **San Joaquin Valley** 圣华金谷是美国加利福尼亚州中央谷地的一片地区，位于萨克拉门托·圣何塞河三角洲南部。

Exercises

I. Read each of the following paraphrases and then write the word it represents on the line provided.

   1 to cause (liquid, for example) to pass through a porous substance or small holes _____

   2 an underground layer of permeable rock, sediment, or soil that yields water _____

   3 to separate and move in different directions _____

   4 capable of being permeated or penetrated, especially by liquids or gases _____

   5 an agent that causes disease, especially a virus, bacterium, or fungus _____

II. Find the Chinese equivalents for the following expressions.

   1 ground water
   2 water table
   3 septic system
   4 surface water
   5 clay layer

III. Discuss the following questions.

   1 What kinds of resources can saltwater provide for humans?
   2 How does surface water percolate into the ground and recharge groundwater?
   3 How can humans access groundwater?
   4 Under what circumstances will ground subsidence occur? What problems can it cause?
   5 How do watershed organizations measure the quality of surface water?

IV. Decide whether the following statements are true or false according to the text.

   ____ 1 Desalination plants are scarce because saltwater is currently not particularly useful when it comes to potable water supplies.

   ____ 2 Most of the pores in the unsaturated zone are filled with water.

_____ 3 All the pores in the saturated zone are filled with water.

_____ 4 An aquifer is "recharged" through precipitation from above percolating through the clay layers.

_____ 5 Groundwater and surface water can recharge each other.

_____ 6 Wells can run dry during times of drought causing the cone of depression to be large.

_____ 7 Saltwater intrusion can occur when the rate of recharging near a shoreline exceeds the rate of pumping.

_____ 8 Subsidence is a temporary phenomenon, but it is extremely costly.

_____ 9 Electrical conductivity is a measure of water quality. So, the more turbid the water, the lower the water quality.

_____ 10 Man-made contaminants such as solvents and pesticides can leach into waterways, degrading the quality of nearby waters.

V. Translate the following sentences into English.

1. 尽管盐水在解决水资源短缺方面于事无补，但确实提供了人类赖以生存的资源。(helpful)
2. 通常，含水层是由具有许多裂缝和相连孔隙的基岩构成的。(be made of)
3. 在大多数情况下，人们将泵放置在井的底部，然后水泵将水抽入房屋，输送到水处理厂。最后，清洁的饮用水从那里分送到千家万户。(disperse)
4. 然而，来自附近污染的土壤和化粪池的污染物会污染水井，使其无法使用。(contaminant)
5. 地面沉降会给地基带来结构性问题。(subsidence)

VI Translate the passage into Chinese.

Aquifers are areas of permeable rock that hold water. Typically, aquifers are made of bedrock that has many fractures and connected pores, such as limestone, sandstone and gravel. Shale and clay layers are impermeable, and therefore make poor aquifers. An aquifer is "recharged" through precipitation from above percolating through the layers of soil and rock. Therefore, there is significant interaction between surface water and groundwater.

Unit 3
Lakes and Rivers

Lead-in

Freshwater on the land surface is a vital part of the water cycle for everyday human life. On the landscape, freshwater is stored in rivers, lakes, reservoirs, creeks and streams. Most of the water people use every day comes from these sources of water on the land surface. To many people, streams and lakes are the most visible part of the water cycle. Not only do they supply the human population, animals, and plants with the freshwater they need to survive, but they are great places for people to have fun.

Text A

Freshwater Lakes and Rivers

1. Rivers and streams are the routes through which fresh water travels to the oceans. Rivers can carry large amounts of water. Earth's largest rivers include the Nile in Africa, the Amazon in South America, and the Mississippi in North America. Thousands of smaller streams, creeks, and brooks carry water into these larger rivers. The fresh water comes from a number of sources, including melting snow, rain, and glacial melt water.

2. Fresh water has a low salt concentration—usually less than 1%. Plants and animals in freshwater regions are adjusted to the low salt content and would not be able to survive in areas of high salt concentration, like the ocean.

3. Lakes can range in size from small ponds to huge bodies of water such as the Great Lakes in the United States. Lakes develop in depressions or basins on Earth's surface. They are filled by water flowing in from streams, from groundwater beneath Earth's surface, and by precipitation and melting snow and ice. Lake water stays fresh as long as water flows in and out of the basin, usually through streams.

4. If there is no outlet, as in the Great Salt Lake in Utah, the water leaves through evaporation. Minerals that are dissolved in the water remain in the lake, and their concentration increases. The lake becomes saline or alkaline.

5. Freshwater rivers and lakes account for approximately 0.009% of Earth's total water and cover a little over 0.3% of the land surface. Lakes and rivers are closely tied. Some lakes are the source for rivers, and some rivers end in lakes. Since both rivers and lakes have fresh water and flow in and out of each other, they share similar characteristics and many plant and animal species.

Rivers and Streams

6 A river is a natural stream of water that flows in one direction in a channel with defined banks. A river or stream forms whenever water moves downhill from one place to another. Some rivers begin high up in the mountains, fed by melting snow or ice. Others collect water from rain and melting snow and ice and from other streams. Others may have their source in springs flowing out of the ground.

7 All rivers flow down a slope. The steeper the slope is, the faster the water flows. As the water reaches flatter areas, it slows down. Water in the Mississippi River moves so slowly that it may appear still. But it is still flowing downhill to the Gulf of Mexico.

8 Rivers flow in channels over different types of ground, or substrate. Water can move into the river channel from the ground on either side of it. The area underground where the water flows is the hyporheic zone. Depending on how gravelly or rocky the substrate is, a river's hyporheic zone can extend far away from the open river channel. The area next to a river channel that is covered with water during a flood is called the floodplain. Some rivers meander through wide valleys with large floodplains. Others may flow swiftly through narrow canyons, with little or no floodplain.

9 Many abiotic factors affect rivers and streams. The temperature of the water depends on where the river is located. Rivers that have their source high in the mountains are fed by melting snow and glaciers. This water is very cold. When it flows through warmer climates, the water is usually warm. But even in a warm place, it is cold if its source is underground.

10 If the river is shallow and flows over a rough substrate like solid rock, the flow will be rough and turbulent. A turbulent river constantly adds oxygen to its water. Deep water flowing slowly over a smooth bottom, such as mud or clay, forms pools. Where deep water flows rapidly over a flat bottom, smooth runs of water are common. Ponds and smooth runs have less oxygen in the water than turbulent areas.

Lakes and Ponds

11 Ponds are small bodies of fresh water shallow enough to support rooted plants. Water temperature is fairly uniform from top to bottom. It changes with air temperature. There is little wave action, because the water is shallow. The bottom of a pond is usually covered with mud. The amount of dissolved oxygen in a pond may vary greatly during a single

day. In very cold locations, an entire pond may freeze solid.

12 Lakes are bigger than ponds and are too deep to support rooted plants except near the shore. Some lakes are deep enough and wide enough for waves to form. Water temperature during summer months is not uniform from top to bottom. Three distinct layers develop.

- The top layer stays warm at around 18.8–24.5°C (65–75°F). This layer is the epilimnion.
- The temperature of the middle layer drops dramatically, usually to 7.4–18.8°C (45–65°F). This layer is the metalimnion.
- The bottom layer is the coldest at around 4.0–7.4°C (39–45°F). This layer is called the hypolimnion.

13 Light does not penetrate to the bottom of the lake, so photosynthesis is limited to the epilimnion. Almost all organisms spend the summer months in the epilimnion with its warmer waters and more abundant food supply.

14 During the fall, overturning occurs. The epilimnion cools with the cooling air and becomes denser than the water below. It sinks and mixes with the hypolimnion. Wind helps mix the water. The lake circulates, or overturns, completely. The lake overturns again in the spring, when the surface ice melts and the resulting cold water sinks.

15 Most lakes are large enough not to freeze solid in winter months. A layer of ice can develop on the top of lakes during winter. The ice blocks out sunlight and can prevent photosynthesis. Without photosynthesis, oxygen levels drop, and some plants and animals may die. This is called winterkill.

16 Because rivers and lakes are often interconnected and water flows between them, many aquatic creatures move between them in search of food. Some creatures prefer to be in either a river or a lake, but many spend some of their lives in both environments.

17 During the winter months some animals hibernate in the quiet bottom mud of rivers and streams. Some fish continue to feed, but less actively.

Rivers and Streams

18 The characteristics of a river or stream change during the journey from its source to its

mouth. The temperature is usually cooler at the source than it is at the mouth. The water is also clearer and has higher oxygen levels. Freshwater fish, such as trout, and heterotrophs can be found nearer the source.

19 As the water flows downstream, the river's width increases and its flow slows down. Species diversity increases with width. Numerous aquatic green plants and algae, which could not live in the faster-moving water upstream, can be found. The nutrient content of the water is largely determined by the land and vegetation surrounding the river. Erosion of the streambed can add to the inorganic nutrients in the water.

20 As water reaches a river's mouth, it becomes murky from the sediments it has picked up. The sediments decrease the amount of light that penetrates the water. With less light, there is less diversity of flora. The river also has less oxygen. Fish living near the mouth of a river, such as catfish and carp, tolerate low levels of oxygen.

21 Many life-forms live in a river's hyporheic zone. Mayflies and other insects and crustaceans can move into the hyporheic zone to hide between rocks. They are protected from predators, yet are in flowing water that brings them nutrients.

Lakes and Ponds

22 Ponds and lakes may have limited diversity of species when they are isolated from one another and from rivers and oceans. Lakes and ponds are divided into three different zones, which are usually determined by depth and distance from the shoreline.

23 The shallow-water, or littoral, zone is the topmost zone near the shore of a lake or pond. It absorbs the Sun's heat and tends to be warmer than deeper zones. Usually a diverse community of organisms lives in the littoral zone. These include several species of algae (like diatoms), rooted and floating aquatic plants, grazing snails, clams, insects, crustaceans, fishes, and amphibians. For some insects, such as dragonflies and midges, only the egg and larval stages are found in this zone. The vegetation and animals living in the littoral zone are food for other creatures such as turtles, snakes, and ducks.

24 The open-water, or limnetic, zone is near-surface open water surrounded by the littoral zone. The limnetic zone is well-lighted like the littoral zone. Both phytoplankton (plants) and zooplankton (animals) thrive in the limnetic zone. Plankton are small, floating

organisms that play a crucial role in the food chain. A variety of freshwater fish also occupy the limnetic zone.

25 Plankton have short life spans. When they die, they sink into the deep water of a lake or pond. This deep-water zone is the profundal zone. This zone is much colder and denser than the other two. Little light penetrates all the way into the profundal zone. Heterotrophs are the organisms that live here. They eat dead organisms and use oxygen for cellular respiration.

(1,521 words)

New Words

1	saline	/ˈseɪlaɪn/ n.	含盐的，咸的
2	alkaline	/ˈælkəlaɪn/ adj.	含碱的，碱性的
3	abiotic	/ˌeɪbaɪˈɒtɪk/ adj.	非生物的
4	substrate	/ˈsʌbstreɪt/ n.	基质
5	hyporheic	/ˌhɪpəˈreɪk/ adj.	潜流的
6	floodplain	/ˈflʌdpleɪn/ n.	漫滩
7	meander	/miˈændə/ n.	缓慢而弯曲地流动
8	turbulent	/ˈtɜːbjələnt/ adj.	湍流的，汹涌的
9	epilimnion	/ˌepɪˈlɪmnɪən/ n.	湖面温水层
10	metalimnion	/ˌmetəˈlɪmnɪən/ n.	变温层，斜温层
11	hypolimnion	/ˌhaɪpəʊˈlɪmnɪən/ n.	下层滞水带，湖底静水层
12	photosynthesis	/ˌfəʊtəʊˈsɪnθəsɪs/ n.	光合作用
13	hibernate	/ˈhaɪbəneɪt/ v.	冬眠
14	trout	/traʊt/ n.	鳟鱼
15	heterotroph	/ˈhetərətrəʊf/ n.	异养生物
16	algae	/ˈældʒiː/ n.	水藻；海藻
17	upstream	/ˌʌpˈstriːm/ adj.	上游的
18	murky	/ˈmɜːki/ adj.	混浊的
19	flora	/ˈflɔːrə/ n.	植物，植物群落

20	mayfly /ˈmeɪflaɪ/ *n.*		蜉蝣
21	crustacean /krʌˈsteɪʃən/ *n. & adj.*		甲壳类动物
22	littoral /ˈlɪtərəl/ *n. & adj.*		海岸的，沿海的
23	larval /ˈlɑːvl/ *n.*		幼虫的，幼虫状态的
24	limnetic /lɪmˈnetɪk/ *adj.*		湖泊的
25	phytoplankton /ˈfaɪtəʊˌplæŋktən/ *n.*		浮游植物
26	zooplankton /ˈzuːəʊˌplæŋktən/ *n.*		浮游动物

Notes

1. **hyporheic zone** 潜流带，是位于溪流或河流河床之下并延伸至河溪边岸带和两侧的水分饱和的沉积物层，地下水和地表水在此交混。潜流带是溪流地表水和地下水相互作用的界面。占据着在地表水、河道之下的可渗透的沉积缓冲带、侧向的河岸带和地下水之间的中心位置。在活跃的河道之下及大多数溪流或河流的河岸带内都可以发现潜流带。在更广泛的意义上，可以把潜流带定义为与地表进行水交换的溪流地下区域。

2. **heterotroph** 异养生物，指的是那些只能将外界环境中现成的有机物作为能量和碳的来源，将这些有机物摄入体内，转变成自身的组成物质，并且储存能量的生物，如：营腐生活和寄生生活的真菌，大多数种类的细菌。

3. **littoral zone** 潮汐带，是指随着潮水的涨落而反复处于海陆交界线附近高潮线和低潮线之间的狭长地带，又称滨海带。其宽度视海岸地形坡度而定。在平缓的海岸地区宽度可达数公里。在坡度很陡的崖岸地区宽度很小。低潮线以下过渡为浅海带，高潮线以上为海岸。

4. **limnetic zone** 湖沼带，潮汐带以外的开阔水域即为湖沼带。湖泊就是由于地势原因积水形成的。沼泽地指长期受积水浸泡，水草茂密的泥泞地区。土壤剖面上部为腐泥沼泽土或泥炭沼泽土，下部为潜育层。有机质含量高，持水性强，透水性弱，干燥时体积收缩。经排水疏干，土壤通气良好，有机物得以分解，可增加肥分。

5. **profundal zone** 深底带，是变温层以下的湖底部分，即营养分解层或湖下层的底部。底质由细微颗粒组成，无任何植物生长。

6. **cellular respiration** 细胞呼吸，是指有机物在细胞内经过一系列的氧化分解，生成无机物或小分子有机物，释放出能量并生成ATP（腺嘌呤核苷三磷酸，是一种不稳定的高能化合物）的过程。

Exercises

I. Read each of the following paraphrases and then write the word it represents on the line provided.

1. of or characterized by the absence of life or living organisms _____
2. to be in a dormant or torpid state during a cold period, especially during the winter _____
3. plants considered as a group, especially the plants of a particular country, region, or time _____
4. moving rapidly or violently _____
5. flat land bordering a river and made up of sediments such as sand, silt, and clay deposited during floods _____

II. Find the Chinese equivalents for the following expressions.

1. littoral zone
2. profundal zone
3. hyporheic zone
4. cellular respiration
5. limnetic zone

III. Discuss the following questions.

1. What abiotic factors can affect rivers and streams?
2. What are the differences between lakes and ponds?
3. What effect do seasons have on the layers of lakes in a year?
4. What is winterkill?
5. How do the biotic characteristics of a river or stream change during the journey from its source to its mouth?
6. Discuss with your classmates about the formation of the Great Salt Lake in Utah.

IV. Choose the best answer to each of the following questions or statements.

1. Depending on how gravelly or rocky the substrate is, a river's _____ zone can extend far away from the open river channel.

    A. profundal   B. hyporheic
    C. limnetic   D. littoral

2 Ponds and smooth runs of water have _____ in the water than turbulent areas.
   A. more oxygen   B. no oxygen
   C. less oxygen   D. the same oxygen

3 Which of the followings is NOT the reason that insects and crustaceans move into the hyporheic zone to hide between rocks?
   A. They are protected from predators.
   B. They are in flowing water that brings them nutrients.
   C. They can use the gaps between the rocks to protect themselves and obtain nutrients.
   D. They can tolerate low levels of oxygen.

4 Ponds and lakes may have limited diversity of species when _____.
   A. they are isolated from one another and from rivers and oceans
   B. they are divided into three different zones
   C. their water temperature during summer months is not uniform from top to bottom
   D. photosynthesis is limited to the epilimnion

5 Both phytoplankton and zooplankton thrive in the _____ zone.
   A. littoral   B. limnetic
   C. hyporheic   D. profundal

V. Translate the following sentences into Chinese.

1 A river is a natural stream of water that flows in one direction in a channel with defined banks.

2 Depending on how gravelly or rocky the substrate is, a river's hyporheic zone can extend far away from the open river channel.

3 Almost all organisms spend the summer months in the epilimnion with its warmer waters and more abundant food supply.

4 Because rivers and lakes are often interconnected and water flows between them, many aquatic creatures move between them in search of food.

5 The nutrient content of the water is largely determined by the land and vegetation surrounding the river.

VI. Translate the following sentences into English.

1 一些河流发源于高山，由融化的冰雪补给水源。(be fed by)

2 一些河流蜿蜒穿过宽阔的山谷。(meander through)

3 湖面温水层随着空气的冷却而冷却，并变得比下面的水密度大。(dense)

4 在冬天的几个月里，一些动物在河底淤泥中冬眠。(hibernate)
5 随着水向下游流动，河流的宽度增加，流速减慢。(flow downstream)

VII. Translate the passage into Chinese.

As water reaches a river's mouth, it becomes murky from the sediments it has picked up. The sediments decrease the amount of light that penetrates the water. With less light, there is less diversity of flora. The river also has less oxygen. Fish living near the mouth of a river, such as catfish and carp, tolerate low levels of oxygen.

Text B

Rivers, Streams, and Creeks

1 Rivers? Streams? Creeks? They are all names for water flowing on the Earth's surface. As far as our Water Science site is concerned, they are pretty much interchangeable. I tend to think of creeks as the smallest of the three, with streams being in the middle, and rivers being the largest.

2 Most of the water you see flowing in rivers comes from precipitation runoff from the land surface alongside the river. Of course, not all runoff ends up in rivers. Some of it evaporates on the journey downslope, can be diverted and used by people for their uses, and can even be lapped up by thirsty animals. Rivers flow through valleys in the landscape with ridges of higher land separating the valleys. The area of land between ridges that collects precipitation is a watershed or drainage basin. Most, but not all, precipitation that falls in a watershed runs off directly into rivers—part of it soaks into the ground to recharge groundwater aquifers, some of which can then seep back into riverbeds.

What is a river?

3 A river forms from water moving from a higher elevation to a lower elevation, all due to gravity. When rain falls on the land, it either seeps into the ground or becomes runoff, which flows downhill into rivers and lakes, on its journey towards the seas. In most landscapes the land is not perfectly flat—it slopes downhill in some direction. Flowing

water finds its way downhill initially as small creeks. As small creeks flow downhill, they merge to form larger streams and rivers. Rivers eventually end up flowing into the oceans. If water flows to a place that is surrounded by higher land on all sides, a lake will form. If people have built a dam to hinder a river's flow, the lake that forms is a reservoir.

Rivers serve many uses

4   The phrase "river of life" is not just a random set of words. Rivers have been essential not only to humans, but to all life on Earth, ever since life began. Plants and animals grow and congregate around rivers simply because water is so essential to all life. It might seem that rivers happen to run through many cities in the world, but it is not that the rivers go through the city, but rather that the city was built and grew up around the river. For humans, rivers are diverted for flood control, irrigation, power generation, public and municipal uses, and even waste disposal.

Where does the water in rivers come from?

5   Large rivers don't start off large at all, but are the result of much smaller tributaries, creeks, and streams combining, just as tiny capillaries in your body merge to form larger blood-carrying arteries and veins.

6   The most simplistic answer is that all the water in a river comes from the sky—and that is certainly true, as streamflow is one part of the water cycle. It is also true that most of the water flowing in rivers comes from precipitation runoff from the surrounding landscape (watershed).

7   But, the water in a river doesn't all come from surface runoff. Rain falling on the land also seeps into the Earth to form groundwater. At a certain depth below the land surface, called the water table, the ground becomes saturated with water. If a river bank happens to cut into this saturated layer, as most rivers do, then water will seep out of the ground into the river. Groundwater seepage can sometimes be seen when water-bearing layers emerge on the land surface, or even on a driveway!

8   The ground below the water table is saturated, whereas the ground above is not. The top layer (unsaturated soil/rock material) is usually wet, but not totally saturated. Saturated, water-bearing materials often exist in horizontal layers beneath the land surface. Since rivers, in time, may cut vertically into the ground as they flow, the water-bearing layers

of rock can become exposed on the river banks. Thus, some of the water in rivers is attributed to flow coming out of the banks. This is why even during droughts there is usually some water in streams.

Where does the water in rivers go?

9   Like everything else on (and in) the Earth, water obeys the rules of gravity and tries to get to the center of the Earth (did you imagine that every molecule in your body is trying to do this, also?). So, the water in rivers flows downhill, with the ultimate goal of flowing into the oceans, which are at sea level. River water may end up in a lake or reservoir, in a pipe aimed at Farmer Joe's corn stalks, in a local swimming pool, or in your drinking glass, but much of it eventually ends up back in the oceans, where it rejoins the water cycle, which is ALWAYS in progress.

10 One river that doesn't participate anymore in this natural water-cycle process is the mighty Colorado River, flowing from Colorado to the Gulf of California in Mexico. The Colorado River certainly starts out as mighty—mighty enough to have carved out the Grand Canyon—but nowadays it does not even end as a trickle. Whereas it once flowed into a large marshy area at the Gulf of California, today it literally disappears in the farmlands at the base of the Sierra de Juarez Mountains in Mexico. As the Colorado River flows through the desert Southwest, it gets used and used by millions of thirsty residents and for crop irrigation; used until there is nothing left.

(921 words)

New Words

1	watershed	/ˈwɔːtəʃed/ n.	分水岭
2	drainage	/ˈdreɪnɪdʒ/ n.	排水
3	elevation	/ˌelɪˈveɪʃən/ n.	海拔，高度
4	seep	/siːp/ v.	渗入，渗透
5	congregate	/ˈkɒŋgrəgeɪt/ v.	聚集
6	municipal	/mjuːˈnɪsɪpəl/ adj.	市政的
7	tributary	/ˈtrɪbjətəri/ n.	支流

| 8 | capillary /kəˈpɪləri/ n. | 毛细管 |
| 9 | trickle /ˈtrɪkəl/ n. | 细流 |

Expressions

1	drainage basin	流域
2	waste disposal	废物处理
3	water-bearing layer	含水层

Notes

1 **drainage basin** 流域，指由分水线所包围的河流集水区。分地面集水区和地下集水区两类。如果地面集水区和地下集水区相重合，称为闭合流域；如果不重合，则称为非闭合流域。平时所称的流域，一般都指地面集水区。
2 **Colorado River** 科罗拉多河，是美国西南方、墨西哥西北方的河流。长度大约有 2333 千米，有一部分在美国落基山脉区域西边消失。整个科罗拉多河河系大部分流入了加利福尼亚湾，是加州淡水的主要来源，另一部分则往南流向墨西哥，但没有出海。

Exercises

I. Read each of the following paraphrases and then write the word it represents on the line provided.

1 the entire region draining into a river, river system, or other body of water _____
2 a thin, irregular, or slow flow of something _____
3 to bring or come together in a group, crowd, or assembly _____
4 a stream that flows into a larger stream or other body of water _____
5 to pass slowly through small openings or pores _____

II. Discuss the following questions.

1 What is a watershed?
2 How is a river formed?
3 Why are rivers essential to all life on Earth?
4 Why is there usually some water in streams even during droughts?
5 Where does the water in rivers go?

III. Decide whether the following statements are true or false according to the text.

　　____ 1　All precipitation runoff from the land surface alongside rivers ends up in rivers.

　　____ 2　Part of precipitation that falls in a watershed soaks into the ground to recharge groundwater aquifers.

　　____ 3　A river forms from water moving from a lower elevation to a higher elevation, all due to gravity.

　　____ 4　In most landscapes the land slopes downhill in some direction.

　　____ 5　If the land surface happens to cut into the water-bearing layer, then water will seep out of the ground into the river.

　　____ 6　Groundwater seepage can be seen when unsaturated rock material emerges on the land surface.

　　____ 7　Saturated water-bearing materials often exist in horizontal layers beneath the land surface.

　　____ 8　Everything on the Earth obeys the rules of gravity and tries to get to the center of the Earth.

　　____ 9　River water may end up in a lake or reservoir, in a pipe, in a local swimming pool, but much of it eventually ends up back in the air.

　　____ 10　Today Colorado River literally flows into the Gulf of California.

IV. Translate the following sentences into English.

　1　一个流域中的大部分降水会直接流入河流。(run off directly into)

　2　当雨水落在地面上时，要么渗入地下，要么变成径流。(seep into)

　3　人们认为河流中的一些水是从河岸渗出的。(be attributed to)

　4　河流中的水顺地势向下流动，最终流入海洋。(flow downhill)

　5　科罗拉多河没有参与自然水循环的过程。(participate in)

V. Translate the passage into Chinese.

But, the water in a river doesn't all come from surface runoff. Rain falling on the land also seeps into the Earth to form groundwater. At a certain depth below the land surface, called the water table, the ground becomes saturated with water. If a river bank happens to cut into this saturated layer, as most rivers do, then water will seep out of the ground into the river. Groundwater seepage can sometimes be seen when water-bearing layers emerge on the land surface, or even on a driveway!

Unit 4
Reservoirs

Lead-in

A reservoir is an artificial lake where water is stored. Most reservoirs are formed by constructing dams across rivers. A reservoir can also be formed from a natural lake whose outlet has been dammed to control the water level. The dam controls the amount of water that flows out of the reservoir.

Text A

Types of Reservoirs

1. The term "reservoir" can refer to a man-made or natural lake, as well as cisterns and subterranean reservoirs. Here we focus only on man-made reservoirs.

2. Man-made reservoirs are made when dams are constructed across rivers, or by enclosing an area that is filled with water. There are two main types of man-made reservoirs: impoundment and off-stream (also called off-river).

3. Reservoirs can vary in size and be as small as a pond and as big as a large lake. There is so much variability when it comes to reservoirs—they can differ in size, shape and location. For this reason, it can be misleading to make blanket statements about reservoirs without "significant qualification as to their type".

4. Depending on the purpose of a reservoir, operators will fill a completed reservoir with water, let water flow on through the dam and downstream, or leave the reservoir site empty until it is needed (e.g. a dry dam site for flood mitigation).

On-stream

5. Storage dams and reservoirs have many uses including drinking water supply. Flood protection, drought management. The design depends on the uses and considerations such as local environmental conditions, geology, topography, a river's flow, fish habitat, and adjacent structures.

    ① Main channel gate opens and closes and can partially block the flow of the river.

    ② Diversion gate channel allows the blocked water (see above) to flow to a storage facility.

    ③ Channel weir, which can also reduce river flow speed, ensures enough water goes toward diversion channel when main channel gate is closed.

Off-stream

6. Reservoirs divert water from a river and temporarily store it in the watershed. Storage

facilities can be reservoirs, wetlands, or even groundwater aquifers.

① Retaining wall holds water in structure.

② Spillway controls release of water when it reaches a certain level.

③ Emergency spillway manages high, flow events by moving extra water downstream.

④ Sluiceway/sluice gate releases water and sediment that collects at the bottom of the toe of a reservoir.

⑤ Hydroelectric generating station produces electricity from flowing water.

Impoundment Reservoir

7   An impoundment reservoir is formed when a dam is constructed across a river. Impoundment reservoirs are usually larger than off-river reservoirs and are the most common form of large reservoirs.

Off-stream Reservoirs

8   Off-stream reservoirs are reservoirs that are not on a river course. Rather, off-stream reservoirs are formed by partially or completely enclosed waterproof banks.

9   The embankments around an off-stream reservoir are usually made from concrete or clay. The size of an off-river reservoir will depend on how large of an area is excavated and how high the embankment is built.

10 Off-stream reservoirs are generally simple in shape and "virtually uniform in depth" compared to impoundment reservoirs, which tend to have shallower shores and varying volumes and shapes.

Construction

11 The construction of a reservoir and dam involves many steps that are taken carefully to ensure the dam is operational and meets all safety standards and environmental regulations.

Diverting the River

12 Before construction can begin, the river must be diverted in order to make the building

process easier. Water can be diverted through constructed channels on the surface alongside the river or through underground tunnels through the rock alongside the river. Both of these methods allow the water to travel downstream of the reservoir site and minimize the amount of water traveling to the construction zone.

13 A temporary dam called a cofferdam is built above the main site of the permanent dam. The cofferdam is intended to protect the construction site in the event of a flood.

14 On wide rivers, a cofferdam may be built on one side of the river to allow water to flow through half of the riverbed. The area behind the cofferdam will be drained and the first half of the main dam will be constructed. This side of the main dam will not be constructed to completion. There will still be some holes in this section of the dam. The cofferdam will then be removed and the water will flow through the holes in the incomplete main dam.

15 Another set of cofferdams is built on the other side of the river and the rest of the main dam is fully constructed. In the final step, the original cofferdam will be reconstructed and the portion of the main dam behind the cofferdam is completed. Once the cofferdam is removed for the last time, the dam is complete and water is stored in the reservoir.

16 In some cases, rather than constructing an on-stream dam, a dam is built off-stream in a topographical depression suitable for holding water (off-stream reservoir). Once the dam is completed, the river will be diverted to the off-stream storage site.

Building the Dam

17 Dam construction depends on the type of dam being built. Arch, gravity and buttress dams are built with concrete and are supported by steel. These dams are used for on-stream storage. These dams used the weight of their structure, abutments, anchors and a solid geological foundation to support them.

18 Embankment dams (either earthfill or rockfill) are made primarily of soil or rock found in or near the construction area to minimize transportation costs. Erosion is a major concern for embankment dams and continuous maintenance such as vegetation control is required. The requirements for the foundation are not as extensive for embankment dams as they are for concrete dams.

Filling the Reservoir

19 Once the dam structure is in place, the reservoir is almost ready to be filled. The land that will be underwater is first surveyed for anything that could potentially contaminate the water. Trash and debris are removed. Then, information signs will be placed around the reservoir and roads leading to the construction area will be barricaded. The reservoir can then be filled.

20 For on-stream storage, water will have already begun collecting behind the dam. However, for off-stream storage, water must be diverted to the reservoir.

21 During the filling process the reservoir site is carefully monitored. Operators will watch for seepage of water through the dam and stay alert for mudslides or landslides, which can occur when the soil and new embankment areas around the filling reservoir are inundated and become wetter than normal.

22 Water quality is monitored at the inflow and outflow of the reservoir and the reservoir water itself. Water quality can be affected by the reservoir since the water that was formerly flowing is now still. Over time, nutrients in the water can build-up and result in the growth of algae blooms (in Alberta, where most reservoirs are located closer to the mountains and inflows often have lower nutrient loads, algae bloom growth is less of a problem than it is in other jurisdictions). Chemicals can also become concentrated in the reservoir, making the water unsuitable for drinking. In Alberta, there are regulations for water quality in a reservoir. The Canadian Fisheries Act regulates the harmful alteration of fish habitat and any required compensation.

Testing Valves and Floodgates

23 Valves are placed in the dam so that water can pass through the dam and go downstream. Valves are tested to ensure that minimum required flows are met and that they can withstand the passage of higher flows if needed. Spillways are common and are often required for safety purposes. A spillway is a part of a dam that allows water to automatically flow over the dam during a flood event.

24 In exceptional cases, when a dam needs to pass flows that are larger than what a valve can pass (e.g. during a flood event), a floodgate is used. These floodgates are tested periodically to ensure that safety standards are met.

Monitoring the Dam

25 The dam will "settle" over time. Since a dam is such a large structure, its height will eventually decrease because its weight will compact it down. The structural integrity of the dam is monitored during this process to ensure that the dam is still functional and safe.

26 Seepage is also monitored. This is particularly important because as the dam ages, tiny micro-fractures can form and the water can begin to pass through the dam wall. Once water begins to pass through, the hole will only get bigger and the problem will become greater.

27 The location of the dam is also monitored. The dam may shift slightly upstream or slightly downstream, which would indicate that the structure is unstable. It can also tilt to one side, which is another sign that the dam is unstable.

28 In addition to monitoring the dam structure, water quality is monitored upstream, in the reservoir and downstream of the dam.

(1,400 words)

New Words

1. cistern /ˈsɪstən/ *n.*   蓄水池
2. subterranean /ˌsʌbtəˈreɪnɪən/ *adj.*   地下的
3. impoundment /ɪmˈpaʊndmənt/ *n.*   蓄水
4. spillway /ˈspɪlweɪ/ *n.*   溢洪道，泄洪道
5. embankment /ɪmˈbæŋkmənt/ *n.*   堤防
6. concrete /ˈkɒŋkriːt/ *n.*   水泥，混凝土
7. clay /kleɪ/ *n.*   黏土
8. excavate /ˈekskəveɪt/ *v.*   挖掘，挖开
9. tunnel /ˈtʌnl/ *n.*   隧道，地道
10. cofferdam /ˈkɒfədæm/ *n.*   围堰
11. riverbed /ˈrɪvəbed/ *n.*   河床
12. drain /dreɪn/ *v.*   排出，排干
13. topographical /ˌtɒpəˈgræfɪkəl/ *adj.*   地形学的，地质的

14	arch /ɑːtʃ/ *n.*	拱，拱状物
15	gravity /ˈɡrævəti/ *n.*	重力
16	buttress /ˈbʌtrəs/ *n.*	拱壁
17	abutment /əˈbʌtmənt/ *n.*	拱基
18	earthfill /ˈɜːθfɪl/ *adj.*	填土的
19	rockfill /ˈrɒkfɪl/ *adj.*	填石的
20	erosion /ɪˈrəʊʒən/ *n.*	腐蚀，侵蚀
21	vegetation /ˌvedʒəˈteɪʃən/ *n.*	植被
22	seepage /ˈsiːpɪdʒ/ *n.*	渗流
23	mudslide /ˈmʌdslaɪd/ *n.*	泥石流
24	jurisdiction /ˌdʒʊərəsˈdɪkʃən/ *n.*	行政辖区
25	valve /vælv/ *n.*	阀门
26	floodgate /ˈflʌdɡeɪt/ *n.*	泄水闸门，防洪闸门
27	integrity /ɪnˈteɡrəti/ *n.*	完整

Expressions

1	subterranean reservoir	地下水库
2	man-made reservoir	人造水库
3	impoundment reservoir	蓄水水库
4	flood mitigation	防洪减灾
5	on-stream reservoir	在槽水库
6	off-stream reservoir	离槽水库
7	waterproof bank	防水堤
8	algae bloom	水华现象

Notes

Algae bloom 水华现象（在海洋中称为"赤潮"）指伴随着浮游生物的骤然大量增殖而直接或间接发生的现象。水面发生变色的情况甚多，厄水（海水变绿褐色）、苦潮（海水变赤色）、青潮（海水变蓝色）及淡水中的水华，都是同样性质的现象。构成水华现象的浮游生物种类很多，但在海洋中鞭毛虫类、硅藻类大多是优势种，在淡水中蓝藻占优势。

Exercises

I. Read each of the following paraphrases and then write the word it represents on the line provided.

1. situated or operating beneath the Earth's surface _____
2. a body of water, such as a reservoir, made by impounding _____
3. a mound of earth or stone built to hold back water or to support a roadway _____
4. to remove by digging or scooping out _____
5. lasting or remaining without essential change _____
6. graphic representation of the surface features of a place or region on a map, indicating their relative positions and elevations _____
7. the natural attraction between physical bodies, especially when one of the bodies is a celestial body, such as the Earth _____
8. the right of a court to hear a particular case, based on the scope of its authority over the type of case and the parties to the case _____
9. something that restrains a flood or outpouring _____
10. the act or process of seeping _____

II. Find the Chinese equivalents for the following expressions.

1. subterranean reservoir
2. man-made reservoir
3. impoundment reservoir
4. on-stream reservoir
5. off-stream reservoir
6. waterproof bank
7. flood mitigation
8. algae bloom

III. Discuss the following questions.

1. How many main types of man-made reservoirs are there? What are they?
2. What steps are involved in reservoir construction?
3. What contributes to algae blooms in the reservoir?
4. After the construction of the dam, regular monitoring becomes a must for the operators. What specific aspects are mentioned with regard to monitoring on a regular basis?

Unit 4   Reservoirs

5   Briefly describe the differences between on-stream reservoirs and off-stream reservoirs.
6   Describe the main steps of reservoir construction to your classmates.

IV. **Choose the best answer to each of the following questions or statements.**

1   Which one of the followings is NOT regarded as a kind of reservoir?
    A. Man-made lake.   B. Natural lake.
    C. Dam.   D. Cistern.

2   What are the three main differences between different reservoirs?
    A. Size, shape and name.   B. Color, shape and location.
    C. Size, name and color.   D. Size, shape and location.

3   What is the top priority for a fresh reservoir construction?
    A. Diverting the river.   B. Building the dam.
    C. Filling the reservoir. D. Testing and monitoring.

4   Constructed with concrete, arch, gravity and buttress dams are sustained by _____.
    A. concrete   B. steel
    C. rock   D. clay

5   Over time, the pace of dam sinking will slow down, and its _____ will at last decrease.
    A. volume   B. mass
    C. weight   D. height

V. **Translate the following sentences into Chinese.**

1   Man-made reservoirs are made when dams are constructed across rivers, or by enclosing an area that is filled with water.
2   In some cases, rather than constructing an on-stream dam, a dam is built off-stream in a topographical depression suitable for holding water (off-stream reservoir).
3   Embankment dams (either earthfill or rockfill) are made primarily of soil or rock found in or near the construction area to minimize transportation costs.
4   In exceptional cases, when a dam needs to pass flows that are larger than what a valve can pass (e.g. during a flood event), a floodgate is used.
5   In addition to monitoring the dam structure, water quality is monitored upstream, in the reservoir and downstream of the dam.

VI. Translate the following sentences into English.
1. "水库"这个术语可以指人工湖泊或天然湖泊，蓄水池和地下水库。(cisterns, subterranean reservoirs)
2. 蓄水水库通常比离槽水库大，属于大型水库中最常见的类型。(impoundment reservoirs)
3. 离槽水库周围的筑堤通常由混凝土或黏土建成。(concrete)
4. 在施工开始前，必须进行河流改道，使施工建设更容易。(divert)
5. 一旦大坝结构成型，水库基本就可以开始准备蓄水工作了。(in place)

VII. Translate the passage into Chinese.

The dam will "settle" over time. Since a dam is such a large structure, its height will eventually decrease because its weight will compact it down. The structural integrity of the dam is monitored during this process to ensure that the dam is still functional and safe.

Seepage is also monitored. This is particularly important because as the dam ages, tiny micro-fractures can form and the water can begin to pass through the dam wall. Once water begins to pass through, the hole will only get bigger and the problem will become greater.

The location of the dam is also monitored. The dam may shift slightly upstream or slightly downstream, which would indicate that the structure is unstable. It can also tilt to one side, which is another sign that the dam is unstable.

Text B

Three Gorges Reservoir Region

1. Three Gorges Reservoir Region, including 25 county-level divisions of Chongqing municipality and Hubei province, is the region directly or indirectly involved in the submersion of the reservoir region of the Three Gorges Dam.

Geographical Condition

2. The Three Gorges Reservoir Region is located in the upstream of the Yangtze River, at the boundary of Chongqing municipality and Hubei province with the area of 59,900 km²

and with the population of 16 million. The Three Gorges Reservoir Region stretches along the Yangtze River from Jiangjin District of Chongqing to Yichang City of Hubei, which is very narrow and where the geography is complex. The mountainous areas represent 74% of the region only with 4.3% plain area in the river valley and 21.7% hilly area. The climate of the reservoir region of the Three Gorges Project is the subtropical monsoon climate. Its location is in the transfer between the northern temperate zone and the subtropical zone, where it is hot and it rains a lot with the annual rainfall of about 1,100 mm. The valley of the Three Gorges Reservoir Region below 500 m has an annual temperature of 17–19 °C, with the frost-free period annually lasts 300–340 days. The annual runoff flow at the site of the dam of the Three Gorges Project is 451 billion m^3 with the annual sediment discharge of 530 million tons.

Natural and Ecological Resources

3   The Three Gorges abounds in water resources, ecological resources, tourism resources, and the resources of some ores, which have a great potential for exploration. The Yangtze River flows through the Three Gorges Reservoir Region with good conditions for water transportation and water-resources development. The potential installed capacity of hydropower can be about 30,000 MW in the Three Gorges Reservoir Region where water resources are the most abundant along the whole main stream of the Yangtze River. Since the Three Gorges Reservoir Region is close to the load center, it has a great potential for the development of water resources. The agricultural resources of the Three Gorges Region are abundant, which are superior for the agriculture, forestry, livestock, sideline production and fishery. The Three Gorges Reservoir Region is proper for the growth of the various plants of both the warm temperate zone and the temperate zone. Dozens of ores have been discovered in the Three Gorges Reservoir Region, such as gold, silver, copper, iron, uranium, sulfur, potassium, natural gas and halite.

4   The natural sights along the Yangtze River in the Three Gorges Reservoir Region are superb, particularly the 162km main river between Sandouping of Yichang and Baidicheng Town of Fengjie, which are the famous sights of the Three Gorges of the Yangtze River and one of the tourist places of interest all over the world. In addition, there are a lot of cultural and historical sites, such as the Qu Yuan Temple of Zigui, the Zhaojun Home Village of Xingshan, the Shibao Stockade of Zhong County and the Ghost Town of Fengdu.

Economy

5   The Three Gorges Reservoir Region is an important part of the economic zone along the Yangtze River. The region is significant for both the promotion to the economic growth of the areas along the Yangtze River and the economic exchanges between the east and the west of China. The region is now and will be more and more superior for the economic progress of China with the gradual opening up to the world and the gradual economic promotion from the coastal areas to the river regions and the inland areas, especially with the construction of the Three Gorges Project. Furthermore, the Three Gorges Region is now and will be more and more significant for the economic progress of China.

6   The Three Gorges Region has made its economic aims as follows.

- To gradually build up the largest hydropower base in China;
- To form an important industrial region midstream and upstream of the Yangtze River;
- To have its particular places of interest for tourism;
- To realize the preliminary economic prosperity;
- To build up the beautiful environments;
- To have a good resettlement of the local people;
- To make a new reservoir region with both the rapid economic progress and the good settlement of the immigrants.

7   During the second phase of reservoir water storage, 25 county-level divisions were involved in the submersion of the reservoir water storage. But from 1992 up to 2002, they had changed a lot as follows.

- Their total GDP had increased 4.1 times, that is, from 8.24 billion RMB yuan to 42.19 billion RMB yuan;
- Their annual growth rate is 17.7%;
- Their annual GDP per head had increased from 906 RMB yuan to 4427 RMB yuan;
- The third industry had properly increased;
- The first industry had largely decreased;

- The second industry had largely increased;

- The percentages of the first, the second and the third industries had changed from 41%, 30% and 29% to 21%, 43% and 36%;

- The total industrial output on the basis of the pricing of 1990 had increased at the rate of 15.7% from 5.64 billion RMB yuan to 24.28 billion RMB yuan, and the total regional industrial output of 2002 was 3.3 times that of 1992;

- The regional revenue had increased at the rate of 16.2% from 790 million RMB yuan to 3.54 billion RMB yuan;

- The annual taxes of the local industrial enterprises had increased at the rate of 19.2% from 420 million RMB yuan to 2.43 billion RMB yuan, and the total annual regional taxes of the local industrial enterprises of 2002 was 4.8 times those of 1992;

- The profits of the local industrial enterprises had changed from -114,500 RMB yuan after the counteraction between the profits and the losses to 720 million RMB yuan after the counteraction between the profits and the losses.

(963 words)

New Words

1	municipality	/mjuːˌnɪsɪˈpælɪti/ n.	直辖市
2	submersion	/səbˈmɜːʃən/ n.	浸没，淹没
3	stretch	/stretʃ/ v.	伸展，延伸
4	subtropical	/ˌsʌbˈtrɒpɪkəl/ adj.	亚热带的
5	monsoon	/mɒnˈsuːn/ n.	季风
6	ore	/ɔː/ n.	矿石，矿物
7	install	/ɪnˈstɔːl/ v.	安装
8	uranium	/jʊˈreɪniəm/ n.	铀
9	sulfur	/ˈsʌlfə/ n.	硫
10	potassium	/pəˈtæsiəm/ n.	钾
11	halite	/ˈhælaɪt/ n.	岩盐
12	counteraction	/ˌkaʊntərˈækʃən/ n.	中和，抵消

Expressions

1	Three Gorges Reservoir	三峡库区
2	the Three Gorges Project	三峡工程
3	subtropical monsoon climate	亚热带季风气候
4	annual rainfall	年降雨量
5	annual temperature	年均气温
6	sideline production	副业生产
7	warm temperate zone	暖温带
8	midstream and upstream of the Yangtze River	长江中上游
9	the third industry	第三产业

Note

GDP (Gross Domestic Product) 国内生产总值，是指按国家市场价格计算的一个国家（或地区）所有常驻单位在一定时期内生产活动的最终成果，常被公认为是衡量国家经济状况的最佳指标。GDP是核算体系中一个重要的综合性统计指标，它反映了一国（或地区）的经济实力和市场规模。

EXERCISES

I. Read each of the following paraphrases and then write the word it represents on the line provided.

1. sinking until covered completely with water _____
2. water, such as rainfall, and any sediment or other substances carried with it that is not absorbed by the soil but instead flows away over the ground _____
3. material that settles to the bottom of a liquid _____
4. to connect or set in position and prepare for use _____
5. a force or influence that makes an opposing force ineffective or less effective _____

II. Find the Chinese equivalents for the following expressions.

1. subtropical monsoon climate
2. annual rainfall
3. annual temperature

4　runoff flow

5　sideline production

6　warm temperate zone

III.　**Discuss the following questions.**

1　What is the location of the Three Gorges Reservoir Region?

2　What are the main natural and ecological resources in the Three Gorges Reservoir Region?

3　Where are the famous sights in the Three Gorges Reservoir Region of the Yangtze River?

4　Why does the Three Gorges Reservoir Region play a significant role in the economic zone along the Yangtze River?

5　During the second phase of reservoir water storage, how many county-level divisions were involved in the submersion of the reservoir water storage?

IV.　**Decide whether the following statements are true or false according to the text.**

_____　1　The reservoir region of the Three Gorges Project enjoys tropical monsoon climate.

_____　2　The annual runoff flow at the site of the dam of the Three Gorges Project is 451 million m^3.

_____　3　The Three Gorges Reservoir Region is appropriate for the growth of various plants.

_____　4　The main river between Sandouping of Yichang and the Ghost Town of Fengdu boasts numerous tourist places of interest.

_____　5　The Three Gorges Reservoir Region is an important part of the economic zone along the Yangtze River.

_____　6　The Three Gorges Region is witnessing its increasingly vital role in the economic progress of China.

_____　7　The Three Gorges Region will gradually build up the largest hydropower base in China.

_____　8　The Three Gorges Region has accomplished its initial economic prosperity.

_____　9　The years from 1992 up to 2002 saw the dramatic changes in the 25 county-level divisions.

_____　10　The third industry had experienced a huge increase in the past twenty years.

V. **Translate the following sentences into English.**

1 三峡库区地处北温带和亚热带的过渡地带,炎热多雨。(transfer)
2 三峡库区具有良好的水运和水资源开发条件。(boast)
3 三峡库区适宜暖温带和温带多种植物的生长。(proper)
4 三峡地区将在中国的经济发展中具有越来越重要的意义。(significant)
5 三峡地区将初步实现经济繁荣。(prosperity)

VI. **Translate the passage into Chinese.**

The Three Gorges Reservoir Region is an important part of the economic zone along the Yangtze River. The region is significant for both the promotion to the economic growth of the areas along the Yangtze River and the economic exchanges between the east of China and the west of China. The region is now and will be more and more superior for the economic progress of China with the gradual opening up to the world and the gradual economic promotion from the coastal areas to the river regions and the inland areas, especially with the construction of the Three Gorges Project.

Unit 5
Flood Control

Lead-in

Flooding is a temporary overflow of water onto land that is normally dry. Floods are the most common natural disaster in many coastal countries. Floods may result from rain, snow, coastal storms, storm surges and overflows of dams and other water systems, and cause outages, disrupt transportation, damage buildings and create landslides.

Text A

Flood Control and Disaster Management

1. Flood control refers to all methods used to reduce or prevent the detrimental effects of flood waters. Some of the common techniques used for flood control are installation of rock berms, rock rip-raps, sandbags, maintaining normal slopes with vegetation or application of soil cements on steeper slopes and construction or expansion of drainage channels. Other methods include levees, dikes, dams, retention or detention basins. After the Katrina Disaster that happened in 2005, some areas prefer not to have levees as flood controls. Communities preferred improvement of drainage structures with detention basins near the sites.

Causes of Floods

2. Floods are caused by many factors: heavy precipitation, severe winds over water, unusual high tides, tsunamis, or failure of dams, levels, retention ponds, or other structures that contained the water.

3. Periodic floods occur on many rivers, forming a surrounding region known as the flood plain.

4. During times of rain or snow, some of the water is retained in ponds or soil, some is absorbed by grass and vegetation, some evaporates, and the rest travels over the land as surface runoff. Floods occur when ponds, lakes, riverbeds, soil, and vegetation cannot absorb all the water. Water then runs off the land in quantities that cannot be carried within stream channels or retained in natural ponds, lakes, and man-made reservoirs. About 30 percent of all precipitation is in the form of runoff and that amount might be increased by water from melting snow. River flooding is often caused by heavy rain, sometimes increased by melting snow. A flood that rises rapidly, with little or no advance warning, is called a flash flood. Flash floods usually result from intense rainfall over a relatively small area, or if the area was already saturated from previous precipitation.

Severe Winds over Water

5   Even when rainfall is relatively light, the shorelines of lakes and bays can be flooded by severe winds—such as during hurricanes—that blow water into the shore areas.

Unusual High Tides

6   Coastal areas are sometimes flooded by unusually high tides, such as spring tides, especially when compounded by high winds and storm surges.

Effects of Floods

7   Flooding has many impacts. It damages property and endangers the lives of humans and other species. Rapid water runoff causes soil erosion and concomitant sediment deposition elsewhere (such as further downstream or down a coast). The spawning grounds for fish and other wildlife habitats can become polluted or completely destroyed. Some prolonged high floods can delay traffic in areas which lack elevated roadways. Floods can interfere with drainage and economic use of lands, such as interfering with farming. Structural damage can occur in bridge abutments, bank lines, sewer lines, and other structures within floodways. Waterway navigation and hydroelectric power are often impaired. Financial losses due to floods are typically millions of dollars each year.

Control of Floods

8   Some methods of flood control have been practiced since ancient times. These methods include planting vegetation to retain extra water, terracing hillsides to slow flow downhill, and the construction of floodways (man-made channels to divert floodwater). Other techniques include the construction of levees, dikes, dams, reservoirs or retention ponds to hold extra water during times of flooding.

Methods of Control

9   In many countries, rivers prone to floods are often carefully managed. Defences such as levees, bunds, reservoirs, and weirs are used to prevent rivers from bursting their banks. When these defences fail, emergency measures such as sandbags or portable inflatable tubes are used. Coastal flooding has been addressed in Europe and the Americas with coastal defences, such as sea walls, beach nourishment, and barrier islands.

10   A dike is another method of flood protection. A dike lowers the risk of having floods compared to other methods. It can help prevent damage; however it is better to combine dikes with other flood control methods to reduce the risk of a collapsed dike.

11   A weir, also known as a low head dam, is most often used to create millponds, but on the Humber River in Toronto, a weir was built near Raymore Drive to prevent a recurrence of the flooding caused by Hurricane Hazel in 1954, which destroyed nearly two fifths of the street.

Flood Clean-up Safety

12   Clean-up activities following floods often pose hazards to workers and volunteers involved in the effort. Potential dangers include electrical hazards, carbon monoxide exposure, musculoskeletal hazards, heat or cold stress, motor vehicle-related dangers, fire, drowning, and exposure to hazardous materials. Because flooded disaster sites are unstable, clean-up workers might encounter sharp jagged debris, biological hazards in the flood water, exposed electrical lines, blood or other body fluids, and animal and human remains. In planning for and reacting to flood disasters, managers provide workers with hard hats, goggles, heavy work gloves, life jackets, and watertight boots with steel toes and insoles.

Benefits of Flooding

13   There are many disruptive effects of flooding on human settlements and economic activities. However, flooding can bring benefits, such as making soil more fertile and providing nutrients in which it is deficient. Periodic flooding was essential to the well-being of ancient communities along the Tigris-Euphrates Rivers, the Nile River, the Indus River, the Ganges and the Yellow River, among others. The viability for hydrologically based renewable sources of energy is higher in flood-prone regions.

(867 words)

Unit 5 Flood Control

New Words

1. berm /bɜːm/ n.   护坡道
2. slope /sləʊp/ n.   斜坡，斜面
3. levee /ˈlevi/ n.   防洪堤
4. dike /daɪk/ n.   堤坝
5. tsunami /tsuːˈnɑːmɪ/ n.   海啸
6. hurricane /ˈhʌrɪkən/ n.   飓风
7. compound /kəmˈpaʊnd/ v.   使恶化
8. concomitant /kənˈkɒmɪtənt/ adj. 相伴的，附随的
9. prolong /prəˈlɒŋ/ v.   延长
10. interfere /ˌɪntəˈfɪə/ v.   干扰，妨碍
11. bund /bʌnd/ n.   堤岸，河岸
12. address /əˈdres/ v.   处理
13. nourishment /ˈnʌrɪʃmənt/ n. 养料，营养
14. collapse /kəˈlæps/ v.   倒塌，崩溃，瓦解
15. weir /wɪə/ n.   堰堤，拦河坝
16. hazard /ˈhæzəd/ n.   危险，危害
17. goggle /ˈɡɒɡəl/ n.   护目镜
18. watertight /ˈwɔːtətaɪt/ adj. 水密的，不漏水的

Expressions

1. storm surge   风暴大潮
2. rock rip-rap   填石
3. drainage channel   排水渠
4. detention basin   滞洪区
5. flood plain   河漫滩
6. surface runoff   地表径流
7. flash flood   山洪暴发
8. sediment deposition 泥沙沉积
9. spawning ground   产卵场
10. bridge abutment   桥台

11	sewer line	污水管道
12	waterway navigation	航道导航
13	hydroelectric power	水力发电
14	carbon monoxide	一氧化碳
15	musculoskeletal hazard	肌肉骨骼危害
16	flood-prone region	洪水易发区

Notes

1 **Hurricane Hazel** 飓风黑兹尔是1954年大西洋飓风季期间造成死亡人数最多、经济损失最惨重的飓风。这场风暴夺走了海地多达1000人的生命，然后又以四级飓风强度从南、北卡罗来纳州边境附近袭击美国，造成95人丧生。之后黑兹尔转变成温带风暴，但仍然致使81人遇难，其中大部分来自多伦多。

2 **Tigris-Euphrates Rivers** 底格里斯—幼发拉底河水系，西南亚最大的河流体系，包含了底格里斯河和幼发拉底河，两河的源头在土耳其东部山间，相距不到80公里，沿东南方向流经叙利亚北部和伊拉克，注入波斯湾。其界定的这一地区的下游以美索不达米亚知名，为文明摇篮之一。幼发拉底河全长约2,800公里，底格里斯河全长约1,900公里。

3 **The Nile River** 尼罗河是一条流经非洲东部与北部的河流，自南向北注入地中海。与中非地区的刚果河以及西非地区的尼日尔河并列非洲最大的三个河流系统。尼罗河长6,670公里，是世界上最长的河流。尼罗河有两条主要的支流，白尼罗河和青尼罗河。发源于埃塞俄比亚高原的青尼罗河是尼罗河下游大多数水和营养的来源，但是白尼罗河则是两条支流中最长的。

4 **The Indus River** 印度河是巴基斯坦主要河流，也是巴基斯坦重要的农业灌溉水源。1947年印巴分治以前，印度河仅次于恒河，为该地区的文化和商业中心地带。河流总长度2900–3200公里。印度河文明为世界上最早进入农业文明和定居社会主要文明之一。

5 **The Ganges** 恒河，中文又译为冈底斯河，为南亚大河，有两个主干，发源于喜马拉雅山南麓和德干高原，流经印度北阿坎德邦、北方邦、比哈尔邦、恰尔肯德邦、西孟加拉邦后，进入孟加拉国，改称帕德玛河，会合布拉马普特拉河在孟加拉国境内的下游贾木纳河，最后注入孟加拉湾，其入海河段称为梅克纳河。

Unit 5 Flood Control

Exercises

I. Read each of the following paraphrases and then write the word it represents on the line provided.

1. a stretch of ground forming a natural or artificial incline _____
2. an enormous sea wave produced by a seaquake or undersea volcanic eruption _____
3. to supply with the maximum that can be held or contained _____
4. to combine so as to form a whole _____
5. the group of natural processes, including weathering, dissolution, abrasion, corrosion, and transportation, by which material is worn away from the Earth's surface _____
6. to be or create a hindrance or obstacle _____
7. to fall down or inward suddenly _____
8. the scattered remains of something broken or destroyed _____
9. a pair of tight-fitting eyeglasses, worn to protect the eyes from hazards such as wind, glare, water, or flying debris _____
10. not permitting the passage of water either in or out; without loopholes. _____

II. Find the Chinese equivalents for the following expressions.

1. rock rip-rap
2. drainage channel
3. detention basin
4. flood plain
5. surface runoff
6. flash flood
7. storm surge
8. sediment deposition
9. spawning ground
10. sewer line

III. Discuss the following questions.

1. What methods are usually adopted to flood control?
2. What are the chief factors causing floods?
3. Why do floods occur during times of rain and snow?
4. What plan can be made to better hold excess water during floods?

5 What measures will be taken if the defences fail in flood-prone countries?

6 Describe the main ways of flood control.

IV. **Choose the best answer to each of the following questions or statements.**

1 What structure is preferable to communities for the sake of flood control?

    A. Levees.   B. Detention basins.

    C. Rock berms.   D. Dikes.

2 What conditions could give rise to flash floods?

    A. Heavy rainfall.   B. Melting snow.

    C. Storm water.   D. Flash flood.

3 Which area is NOT mentioned in terms of the influence of floods?

    A. Manufacturing.   B. Agriculture.

    C. Traffic.   D. Hydroelectric power.

4 The most common method of flood control is _____.

    A. piling up sandbags   B. using portable inflatable tubes

    C. building barrier islands   D. building defences

5 Which materials are NOT supplied to the cleaning workers for dealing with flood disasters after the flood?

    A. Hard hats.   B. Goggles.

    C. Normal gloves.   D. Watertight boots.

V. **Translate the following sentences into Chinese.**

1 Flood control refers to all methods used to reduce or prevent the detrimental effects of flood waters.

2 Periodic floods occur on many rivers, forming a surrounding region known as the flood plain.

3 Flash floods usually result from intense rainfall over a relatively small area, or if the area was already saturated from previous precipitation.

4 Coastal areas are sometimes flooded by unusually high tides, such as spring tides, especially when compounded by high winds and storm surges.

5 The viability for hydrologically based renewable sources of energy is higher in flood-prone regions.

VI. Translate the following sentences into English.
1 社区倾向于改善排水结构，在现场附近设置滞洪区。(drainage, detention basin)
2 河流洪水通常是由暴雨引起的，有时由于积雪融化而加剧。(increase)
3 雨水径流会快速导致其他地方的泥土流失和与之相伴的沉积物的沉积。(erosion, sediment deposition)
4 最好将堤防与其他防洪方法结合起来，以减少堤防倒塌的风险。(collapse)
5 洪水过后的清理活动经常对参与清理工作的工人和志愿者造成危害。(pose hazard)

VII. Translate the passage into Chinese.

Flooding has many impacts. It damages property and endangers the lives of humans and other species. Rapid water runoff causes soil erosion and concomitant sediment deposition elsewhere (such as further downstream or down a coast). The spawning grounds for fish and other wildlife habitats can become polluted or completely destroyed. Some prolonged high floods can delay traffic in areas which lack elevated roadways. Floods can interfere with drainage and economic use of lands, such as interfering with farming. Structural damage can occur in bridge abutments, bank lines, sewer lines, and other structures within floodways. Waterway navigation and hydroelectric power are often impaired. Financial losses due to floods are typically millions of dollars each year.

Text B

More Floods, More Effective Flood-fighting Technology

1 As floods become more severe and more frequent, government authorities must invest in advanced technology platforms that take the guesswork out of crisis management. Since the consequences of flood events vary dramatically, the tools used to fight them—such as surveillance, connectivity, and situational awareness technologies—must be able to adapt to each unique situation.

2 In 2018 alone, there were 14 weather-related natural disasters across the United States that caused total damage exceeding $65 billion. From hurricanes and wildfires to droughts

and floods, natural disasters have had devastating effects in recent years, with increasing frequency. Although all of these threats deserve renewed attention from government authorities and first responders, floods have been especially destructive, from the deadly 2005 flooding in the wake of Hurricane Katrina to the recent flooding in the Midwest.

3 As floods take a greater toll on the nation's aging national infrastructure system and put communities across the country at risk, it is incumbent on government authorities to develop flood preparedness and response plans equal to the magnitude of the challenge. To do so, local, state, and federal authorities must surmount historical obstacles that have (perhaps until recently) largely stymied governmental agencies and first responders alike. Many of these lessons have been learned the hard way, but they underscore the importance of "getting it right" in the immediate aftermath of major flooding events.

Flooding: A Rising Threat

4 Natural disasters of all kinds pose challenges for communities across the country. However, flooding—the end result of events such as hurricanes and late winter storms—carries significant consequences in both the short- and long-term. In addition to the 2019 flooding in the Dakotas, Nebraska, and Montana (events that have already pushed local governments to the breaking point), the past decade has witnessed floods that have left indelible marks on communities and resulted in the loss of human life. Many of these events have also posed serious threats to national security and defense apparatus.

5 The floods that accompanied Hurricane Harvey paralyzed one of the nation's largest metropolitan areas and contributed to the deaths of nearly 100 people. Superstorm Sandy, which struck New York and New Jersey in 2012 and resulted in more than $60 billion worth of damage, resulted in floods so severe that nearly half of all deaths caused by the disaster were related to flooding.

6 For government authorities, which marshal disaster relief resources and organize crisis management, flooding poses unique challenges. Often simultaneously affecting broad geographic areas ranging from urban zones to rural communities, floods demand synchronized action on multiple fronts. Additionally, the essential components of orderly and unified disaster relief (e.g., effective information sharing and community outreach) may become even more strained during floods, which can cause power outages, interfere with Internet and cellular connectivity, and disable phone lines.

The Role of Technology in Disaster Management

7 Although the short- and long-term consequences of increasingly frequent flooding vary from one community to the next, what is clear is that authorities across multiple levels of government need to invest in smart, strategic flood-response technology to prepare disaster response teams for the immediate aftermath of these events. The most effective flood response technologies range from drones for surveillance to situational awareness platforms that ensure valuable data reaches the right people. These tools allow decision-makers to build a comprehensive plan of action and respond to flooding events in real-time.

8 Although natural disasters have always presented considerable challenges to authorities—especially at the local level where resources and personnel can be limited—flood responses in particular demand a significant degree of coordination and collaboration. By incorporating surveillance, connectivity, and situational awareness technologies, it should be possible to launch emergency responses that are faster, better organized, and ultimately more effective in mitigating the damage of flood events.

9 It is essential that emergency response teams deployed to flooded areas operate from a shared common operational picture (COP)—one that provides reliable, real-time information as to the whereabouts of citizens, response crews, and infrastructural assets. In other words, the COP should enable high-fidelity situational awareness across all relevant personnel. By working from sound intelligence, disaster management efforts can be organized from the top down, rather than by way of the traditional "every crew for itself" mentality. At the same time, flood-response technology can enable and encourage personnel in the field to share important information with other deployed personnel and with a central headquarters.

Emerging Flood-response Technologies

10 Recent floods have underscored the importance of communication and coordination in developing real-time responses. Fortunately, the technologies that are emerging are imminently capable of adding tremendous value on a number of fronts.

Pivotal flood-response technologies include:

- *Surveillance Technology*—Drones or unmanned aerial vehicles (UAVs) provide ground crews with an "eye in the sky" that allows them to gain a full picture of the flooding, including which areas are most affected and where emergency responses may be required next. Underwater drones can also help responders examine infrastructure and coordinate rescue efforts in heavily flooded areas.

- *Flood Mapping Technology*—Flood forecast maps use remote sensors to determine which areas are most at risk of flooding based on elevation, proximity to bodies of water, and other topographical data. They can also be helpful in evaluating if and when to rebuild infrastructure after a natural disaster, as some areas may have become too dangerous to accommodate homes and businesses.

- *Connectivity Technology*—The most powerful tool available during a natural disaster is one that nearly everyone has access to: a smartphone. Quickly deployable cellular data communication platforms can help people stay in contact with their loved ones during floods. These networks can also help authorities communicate more easily with one another and with imperiled communities over social media.

- *Situational Awareness Technology*—Situational awareness platforms integrate discrete technologies, synthesize information streams, and activate data from UAVs, intelligent infrastructure, meteorological data, and more. Such integration helps responders build a COP, enabling crews on the ground to execute their critical responsibilities with far greater effectiveness.

11 A new generation of digital technologies is helping governments take decisive control over major flooding events. Perhaps the most important development in flood-response technology is the rise of situational awareness platforms. This technology enables decision-makers to effectively coordinate response efforts at a moment's notice, rather than executing strategies designed for dynamic situations that will almost certainly have changed by the time first responders arrive on the scene.

12 Integrating flood-response technologies and disaster response personnel into a situational awareness platform can make real-time coordination a reality, and help prioritize and distribute mission-critical information to the right people at the right time. It is even possible to leverage the power of crowdsourcing to pull citizen-generated data from social media and purpose-built public applications into a COP. Through the use of mobile data communication platforms, citizens are able to support the COP and receive vital information from government responders.

13 Ultimately, surveillance, connectivity, and situational awareness technologies have the potential to revolutionize how governments respond to major flooding events. By leveraging information in a more coordinated fashion and pulling from a wide array of assets across disaster-stricken areas, it is possible to develop responses that are better organized in turn—and save lives in the process.

(1,189 words)

New Words

1	guesswork /ˈɡeswɜːk/	*n.*	臆测，猜测
2	surveillance /səˈveɪləns/	*n.*	监视，监督
3	devastating /ˈdevəsteɪtɪŋ/	*adj.*	毁灭性的
4	responder /rɪˈspɒndə/	*n.*	响应者
5	incumbent /ɪnˈkʌmbənt/	*adj.*	有责任的
6	surmount /səˈmaʊnt/	*v.*	克服
7	stymy /ˈstaɪmi/	*v.*	阻挠，妨碍
8	underscore /ˌʌndəˈskɔː/	*v.*	强调
9	aftermath /ˈɑːftəmæθ/	*n.*	后果，余波
10	indelible /ɪnˈdeləbəl/	*adj.*	不能擦除的，不能磨灭的
11	metropolitan /ˌmetrəˈpɒlɪtən/	*adj.*	大都市的
12	marshal /ˈmɑːʃəl/	*v.*	整顿，配置
13	synchronize /ˈsɪŋkrənaɪz/	*v.*	使同步
14	drone /drəʊn/	*n.*	无人驾驶飞机
15	mitigate /ˈmɪtɪɡeɪt/	*v.*	减轻，缓和
16	mentality /menˈtæləti/	*n.*	心态，思想方法
17	imminently /ˈɪmɪnəntli/	*adv.*	迫切地，紧急地
18	pivotal /ˈpɪvətəl/	*adj.*	重要的，关键的
19	aerial /ˈeəriəl/	*adj.*	空中的
20	proximity /prɒkˈsɪməti/	*n.*	接近
21	discrete /dɪˈskriːt/	*adj.*	不连续的，分离的
22	synthesize /ˈsɪnθəsaɪz/	*v.*	合成，综合

23	activate /ˈæktɪveɪt/ v.	激活
24	meteorological /ˌmiːtiərəˈlɒdʒɪkəl/ adj.	气象的
25	prioritize /praɪˈɒrɪtaɪz/ v.	按优先顺序处理
26	leverage /ˈliːvərɪdʒ/ v.	利用
27	crowdsource /ˈkraʊdsɔːs/ n.	众包（指把工作任务外包给大众的做法）

Expressions

1	situational awareness	情景感知
2	take a toll	造成损失；产生负面影响
3	breaking point	（问题难以遏制的）顶点，极限
4	defense apparatus	防御设施
5	power outage	停电
6	flood-response technology	洪水响应技术
7	real-time	实时的
8	common operational picture	共用作战图
9	high-fidelity	高保真
10	unmanned aerial vehicle	无人机
11	underwater drone	水下无人机
12	dynamic situation	动态情景

Notes

1. **Hurricane Katrina** 飓风卡特里娜于2005年8月中在巴哈马群岛附近生成，于佛罗里达州以小型飓风强度登陆。随后数小时，该风暴进入了墨西哥湾，迅速增强为5级飓风。卡特里娜飓风整体造成的经济损失可能高达2,000亿美元，至少有1,833人丧生，成为美国史上破坏最大的飓风。

2. **Hurricane Harvey** 飓风哈维是2017年大西洋飓风季中的一个热带气旋。2017年8月25日，登陆美国德克萨斯州沿岸，时速130英里。

Exercises

I. Read each of the following paraphrases and then write the word it represents on the line provided.

1. the systematic observation of aerospace, surface, or subsurface areas, places, persons, or things, by visual, aural, electronic, photographic, or other means _____
2. a remotely controlled or autonomous aircraft with no pilot on board _____
3. to make less severe or intense; moderate or alleviate _____
4. existing, occurring, moving, or operating in the air _____
5. nearness in space or time _____
6. to cause to appear important or deserving of attention _____
7. to combine or cause to combine into a whole _____
8. to set in motion; make active or more active _____
9. being of vital or central importance; crucial _____
10. to arrange or deal with in order of importance _____

II. Find the Chinese equivalents for the following expressions.

1. situational awareness
2. common operational picture
3. high-fidelity
4. dynamic situation
5. defense apparatus
6. flood-response technology
7. common operational picture
8. real-time
9. high-fidelity
10. unmanned aerial vehicle

III. Discuss the following questions.

1. What are main tools used to fight floods?
2. In the author's opinion, what do authorities of all levels need to do to get prepared for disaster response?
3. What are the most effective flood response technologies about?
4. What are the functions of situational awareness technology?
5. What can governments do with the help of situational awareness platforms?

IV. Decide whether the following statements are true or false according to the text.

____ 1 Floods can cause power outages, interfere with Internet and cellular connectivity, and disable phone lines.

____ 2 By working from sound intelligence, disaster management efforts can be organized by way of "every crew for itself" mentality.

____ 3 Flood-response technology can enable flood fighting personnel to gain a full picture of the flooding area.

____ 4 Surveillance technology can allow ground crews to share important information with a central headquarters.

____ 5 Governments' response to major flooding events can be revolutionized with the help of surveillance, connectivity, and situational awareness technologies.

V. Translate the following sentences into English.

1 洪灾会对国家安全和人民的生命构成严重威胁。(pose threats)
2 洪灾会造成停电，干扰互联网和手机通讯，使电话线瘫痪。(interfere with)
3 洪水响应技术可以使现场人员与其他部署的人员共享重要信息。(enable)
4 新一代的数字技术正在帮助政府对重大洪水事件进行决定性的控制。(take control over)
5 监视、通讯和情景感知技术有可能彻底改变政府应对重大洪水事件的方式。(revolutionize)

VI. Translate the passage into Chinese.

As floods become more severe and more frequent, government authorities must invest in advanced technology platforms that take the guesswork out of crisis management. Since the consequences of flood events vary dramatically, the tools used to fight them — such as surveillance, connectivity, and situational awareness technologies — must be able to adapt to each unique situation.

Unit 6

Irrigation and Drainage

Lead-in

Irrigation is the artificial application of water to land and drainage is the artificial removal of excess water from land. Some land requires irrigation or drainage before it is possible to use it for any agricultural production; other land profits from both practices to increase production. Some land, of course, does not need either of them. Irrigation and drainage improvements are not necessarily exclusive. Often both may be required together to assure sustained, high-level production of crops.

Text A

Modern Irrigation System Planning and Construction

1. The first consideration in planning an irrigation project is developing a water supply. Water supplies may be classified as surface or subsurface. Though both surface and subsurface water come from precipitation such as rain or snow, it is far more difficult to determine the origin of subsurface water.

2. In planning a surface water supply, extensive studies must be made of the flow in the stream or river that will be used. If the streamflow has been measured regularly over a long period, including times of drought and flood, the studies are greatly simplified. From streamflow data, determinations can be made of the minimum, maximum, average daily, and average monthly flows; the size of dams, spillways, and downstream channel; and the seasonal and carry-over storage needed. If adequate streamflow data are not available, the streamflow may be estimated from rain and snow data, or from flow data from nearby streams that have similar climatic and physiographic conditions.

3. The quality, as well as the quantity, of surface water is a factor. The two most important considerations are the amount of silt carried and the kind and amount of salts dissolved in the water. If the silt content is high, sediment will be deposited in the reservoir, increasing maintenance costs and decreasing useful life periods. If the salt concentration is high, it may damage crops or accumulate in the soil and eventually render it unproductive.

4. Subsurface sources of water must be as carefully investigated as surface sources. In general, less is known about subsurface supplies of water than about surface supplies, so, therefore, subsurface supplies are harder to investigate. Engineers planning a project need to know the extent of the basic geological source of water (the aquifer), as well as the amount the water level is lowered by pumping and the rate of recharge of the aquifer. Often the only way for the engineer to obtain these data reliably is to drill test wells and make on-site measurements. Ideally, a project is planned so as not to use more subsurface water than is recharged. Otherwise, the water is said to be "mined," meaning that it is

being used up as a natural resource and its use is considered unsustainable.

5   Two sources of water not often thought of by the layperson are dew and sewage or wastewater. In certain parts of the world, Israel and part of Australia, for example, where atmospheric conditions are right, sufficient dew may be trapped at night to provide water for irrigation. Elsewhere the supply of wastewater from some industries and municipalities is sufficient to irrigate relatively small acreages. Recently, due to greater emphasis on purer water in streams, there has been increased interest in this latter practice.

6   In some countries (Egypt for example), sewage is a valuable source of water. In others, such as the United States, irrigation is looked upon as a means of disposing of sewer water as a final step in the wastewater treatment process. Unless the water contains unusual chemical salts, such as sodium, it is generally of satisfactory quality for agricultural irrigation. Where the practice is used primarily as a means of disposal, large areas are involved and the choice of crop is critical. Usually only grass or trees can withstand the year-round applications.

7   Before a water supply can be assured, the right to it must be determined. Countries and states have widely varying laws and customs that determine ownership of water. If the development of a water supply is for a single purpose, then the determination of ownership may be relatively simple; but if the development is multipurpose, as most modern developments are, ownership may be difficult to determine, and agreements must be worked out among countries, states, municipalities, and private owners.

8   The area that can be irrigated by a water supply depends on the weather, the type of crop grown, and the soil. Numerous methods have been developed to evaluate these factors and predict average annual volume of rainfall needed. Some representative annual amounts of rainfall needed for cropland in the western United States are 305 to 760 mm (12 to 30 inches) for cereal grains and 610 to 1,525 mm (24 to 60 inches) for forage. In the Near East, cotton needs about 915 mm (36 inches), whereas rice may require two to three times that amount. In humid regions of the United States, where irrigation supplements rainfall, grain crops may require 150 to 230 mm (6 to 9 inches) of water. In addition to satisfying the needs of the crop, allowances must be made for water lost directly to evaporation and during transport to the fields.

Transport Systems

9 The type of transport system used for an irrigation project is often determined by the source of the water supply. If a surface water supply is used, a large canal or pipeline system is usually required to carry the water to the farms because the reservoir is likely to be distant from the point of use. If subsurface water drawn from wells is used, a much smaller transport system is needed, though canals or pipelines may be used. The transport system will depend as far as possible on gravity flow, supplemented if necessary by pumping. From the mains, water flows into branches, or laterals, and finally to distributors that serve groups of farms. Many auxiliary structures are required, including weirs (flow-diversion dams), sluices, and other types of dams. Canals are normally lined with concrete to prevent seepage losses, control weed growth, eliminate erosion hazards, and reduce maintenance. The most common type of concrete canal construction is by slip forming. In this type of construction, the canal is excavated to the exact cross section desired and the concrete placed on the earth sides and bottom.

10 Pipelines may be constructed of many types of material. The larger lines are usually concrete whereas laterals may be concrete, cement-asbestos, rigid plastic, aluminum, or steel. Although pipelines are more costly than open conduits, they do not require land after construction, suffer little evaporation loss, and are not troubled by algae growth.

Water Application

11 After water reaches the farm, it may be applied by surface, subsurface, or sprinkler-irrigation methods. Surface irrigation is normally used only where the land has been graded so that uniform slopes exist. Land grading is not necessary for other methods. Each method includes several variations, only the more common of which are considered here.

12 Surface irrigation systems are usually classed as either flood or furrow systems. In the flood system, water is applied at the edge of a field and allowed to move over the entire surface to the opposite side of the field. Grain and forage crops are quite often irrigated by flood techniques.

13 The furrow system is used for row crops, such as corn (maize), cotton, sugar beets, and potatoes. Furrows are plowed between crop rows and the water is run in the furrows. In either type of surface irrigation systems, waste-water ditches at the lower edge of the

fields permit excess water to be removed for use elsewhere and to prevent waterlogging.

14 Subirrigation is a less common method. An impermeable layer must be located below, but near the root zone of the crop so that water is trapped in the root zone. If this condition exists, water is applied to the soil through tile drains or ditches.

15 Sprinklers have been used increasingly to irrigate agricultural land. Little or no preparation is needed, application rates can be controlled, and the system may be used for frost protection and the application of chemicals, such as pesticides, herbicides, and fertilizers. Sprinklers range from those that apply water in the form of a mist to those that apply an inch or more per hour.

(1,280 words)

New Words

1. irrigation /ˌɪrɪˈgeɪʃən/ n.　　灌溉
2. streamflow /ˈstriːmfləu/ n.　　流速及流水量
3. physiographic /ˈfɪzɪəˈgræfɪk/ adj.　　地形学的
4. render /ˈrendə/ v.　　导致
5. aquifer /ˈækwɪfə/ n.　　含水土层
6. drill /drɪl/ v.　　钻孔
7. mine /maɪn/ v.　　开采
8. layperson /ˈleɪˌpɜːsən/ n.　　外行，非专业人员
9. dew /djuː/ n.　　露水
10. sewage /ˈsjuːɪdʒ/ n.　　污水
11. acreage /ˈeɪkərɪdʒ/ n.　　大块土地
12. multipurpose /ˌmʌltɪˈpɜːpəs/ adj.　　多种用途的，多目标的
13. forage /ˈfɒrɪdʒ/ n.　　草料，饲料
14. lateral /ˈlætərəl/ n.　　侧面
15. auxiliary /ɔːgˈzɪljəri/ adj.　　辅助的，附加的
16. aluminum /ˌæljəˈmɪnəm/ n.　　铝
17. sprinkler /ˈsprɪŋklə/ n.　　洒水装置，洒水车

18	furrow /ˈfʌrəʊ/ n.	畦沟
19	waterlog /ˈwɔːtəlɒg/ v.	使（船等）进水，使浸透水
20	subirrigation /ˌsʌbɪrɪˈgeɪʃn/ n.	地下灌溉（通过地下多孔管道灌溉）

Expressions

1	irrigation project	灌溉工程
2	subsurface water	地下水
3	silt content	含泥沙量
4	on-site measurement	现场测量
5	wastewater treatment	废水处理
6	agricultural irrigation	农业灌溉
7	slip form	滑模
8	rigid plastic	硬质塑料
9	sprinkler-irrigation	喷灌
10	impermeable layer	不透水膜

Note

slip form 滑模。滑模工程技术是我国现浇混凝土结构工程施工中机械化程度高、施工速度快、现场场地占用少、结构整体性强、抗震性能好、安全作业有保障、环境与经济综合效益显著的一种施工技术，通常简称为"滑模"。

Exercises

I. Read each of the following paraphrases and then write the word it represents on the line provided.

1 supplying dry land with water by means of ditches _____

2 one who is a nonprofessional in a given field _____

3 water droplets condensed from the air, usually at night, onto cool surfaces _____

4 liquid and solid waste carried off in sewers or drains _____

5 designed or used for several purposes _____

6 of, relating to, or situated at or on the side _____

7 plant material that livestock graze or that is cut and fed to them _____
8 a long narrow trench or furrow dug in the ground, as for irrigation, drainage, or a boundary line _____
9 to saturate with water and make soggy or unusable _____
10 a substance used to destroy or inhibit the growth of plants, especially weeds _____

II. **Find the Chinese equivalents for the following expressions.**
1 subsurface water
2 irrigation project
3 silt content
4 on-site measurement
5 wastewater treatment
6 agricultural irrigation
7 slip form
8 rigid plastic
9 sprinkler-irrigation
10 impermeable layer

III. **Discuss the following questions.**
1 What do the two types of water supplies refer to? What are their main sources?
2 What are the top considerations that affect the quality and quantity of surface water?
3 How do engineers who plan a project get data on the basic geological sources of water?
4 What are the two types of surface irrigation systems?
5 What are the advantages of sprinklers?
6 Describe the transport system of an irrigation project.

IV. **Choose the best answer to each of the following questions or statements.**
1 What is the top priority in planning an irrigation project?
   A. Developing a water supply.
   B. Investigating the subsurface sources of water.
   C. Assuring water supply right.
   D. Improving soil conditions.

2 The influencing factors of water supply irrigation areas do NOT include _____.
   A. the climate   B. the weather
   C. the type of crop grown   D. the soil

3 Which means could NOT be employed in terms of water application?
   A. Surface irrigation.   B. Agricultural irrigation.
   C. Subsurface irrigation.   D. Sprinkler-irrigation.

4 Which crop is more suitably irrigated by flood techniques?
   A. Corn.   B. Cotton.
   C. Forage crop.   D. Sugar beet.

5 What chemicals can NOT be applied to sprinklers?
   A. Pesticide.   B. Herbicide.
   C. Fertilizer.   D. Sodium salt.

V. **Translate the following sentences into Chinese.**

1 Irrigation is looked upon as a means of disposing of sewer water as a final step in the wastewater treatment process.

2 Before a water supply can be assured, the right to it must be determined.

3 In addition to satisfying the needs of the crop, allowances must be made for water lost directly to evaporation and during transport to the fields.

4 In this type of construction, the canal is excavated to the exact cross section desired and the concrete placed on the earth sides and bottom.

5 After water reaches the farm it may be applied by surface, subsurface, or sprinkler-irrigation methods.

VI. **Translate the following sentences into English.**

1 非专业人员常忽略两种水源：露水和污水（或废水）。(layperson)

2 最近，由于人们更加重视更为纯净的溪流水，所以对废水处理越来越感兴趣。(emphasis on)

3 如果水中不含非常见的化学盐，用其进行农业灌溉效果很好。(agricultural irrigation)

4 犁沟法用于种植玉米、棉花、甜菜和马铃薯等行播作物。(furrow system)

5 塑料长期暴露在阳光下容易老化。(expose)

VII. **Translate the passage into Chinese.**

The type of transport system used for an irrigation project is often determined by the source of the water supply. If a surface water supply is used, a large canal or pipeline system is usually required to carry the water to the farms because the reservoir is likely to be distant from the point of use. If subsurface water drawn from wells is used, a much smaller transport system is needed, though canals or pipelines may be used. The transport system will depend as far as possible on gravity flow, supplemented if necessary by pumping.

Text B

Modern Drainage System Construction

Types of Drainage Systems

1. Drainage systems may be divided into two categories, surface and subsurface. Each has several components with similar functions but different names. At the lower, or disposal end of either system is an outlet. In order of decreasing size, the components of a surface system are the main collecting ditch, field ditch, and field drain; and for a subsurface system, main, submain, and lateral conduits from the submain. The outlet is the point of disposal of water from the system; the main carries water to the outlet; the submain or field ditch collects water from a number of smaller units and carries it to the main; and the lateral or field drain, the smallest unit of the system, removes the water from the soil.

2. The outlet for a drainage system may be a natural stream or river or a large constructed ditch. A constructed ditch usually is trapezoidal in section with side banks flat enough to be stable. Grass may be grown on the banks, which are kept clear of trees and brush that would interfere with the flow of water.

3. A surface drainage system removes water from the surface of the soil and to approximately the bottom of the field ditches. A surface system is the only means for drainage improvement on soils that transmit water slowly. Individual surface drains also are used to supplement subsurface systems by removing water from ponded areas.

4. The field drains of a surface system may be arranged in many patterns. Probably the two most widely used are parallel drains and random drains. Parallel drains are channels running parallel to one another at a uniform spacing of a few to several hundred metres apart,

depending on the soil and the slope of the land. Random drains are channels that run to any low areas in the field. The parallel system provides uniform drainage, whereas the random system drains only the low areas connected by channels. In either case the channels are shallow with flat sides and may be farmed like the rest of the field. Crops are usually planted perpendicular to the channels so that the water flows between the rows to the channels.

5   Some land grading of the fields where surface drains are installed is usually essential for satisfactory functioning. Land grading is the shaping of the field so that the land slopes toward the drainage channels. The slope may be uniform over the entire field or it may vary from part to part. Historically, the calculations necessary for planning land grading were time-consuming, a factor that restricted the alternatives available for final design. Today, computer models rapidly explore many possibilities before a final land grading design is selected.

6   In a subsurface drainage system, often called a tile system, all parts except the outlet are located below the surface of the ground. It provides better drainage than a surface system because it removes water from the soil to the depth of the drain, providing plants a greater mass of soil for root development, permitting the soil to warm up faster in the spring, and maintaining a better balance of bacterial action, the air in the soil, and other factors needed for maximum crop growth.

7   The smallest component of the subsurface system, the lateral, primarily removes water from the soil. The laterals may be arranged in either a uniform or a random pattern. The choice is governed by the crop grown and its value, the characteristics of the soil, and the precipitation pattern.

8   The primary decision required for a system with uniform laterals is their depth and spacing. In general, the deeper the laterals can be emplaced, the farther apart they can be spaced for an equivalent degree of drainage. Laterals usually are spaced from 24 to 91 meters (80 to 300 feet) apart and 0.9 to 1.5 meters (3 to 5 feet) deep.

9   Subsurface drainage systems are as important in many irrigated areas as they are in humid areas. A drainage system is needed on irrigated lands to control the water table and ensure that water will be able to move through a soil, thus keeping salts from accumulating in the root zone and making the soil unproductive.

Construction and Maintenance

10 Most subsurface drains are constructed by excavating a trench, installing a tile, and backfilling the trench. Control of the machines to assure proper slope of the drain has been a major problem, but recent development in excavation technology, including the use of laser beams for grade control, has helped to solve it. Traditionally, clay or concrete tile has been the principal material used, but many types of perforated plastic tubes are now employed. An advantage is the reduction in weight of the material handled.

11 With proper maintenance, drainage systems give relatively long life. Selected herbicides are often applied to keep woody growth and waterweed out of the channels. Grates are usually installed over outlets to prevent rodents and burrowing animals from building nests.

12 Surface drainage systems need almost yearly maintenance to assure the slope and cross section of the channels and the slope of the graded areas because the slopes are so flat that small changes in the ground surface can make marked changes in the ability of a system to function.

13 A combination of drainage and irrigation is used to reclaim large areas of land that have been abandoned because of salt accumulation. In this case subsurface drainage systems must be installed so that high water tables are lowered and pure water flushed through the soil, dissolving the salts and carrying them away in the drainage water. Large areas in the United States, India, and the Middle East are potentially available for reclamation by this technique.

14 The people of the Netherlands have reclaimed land from the sea by the use of drainage. Since the IJsselmeer (formerly Zuiderzee) barrier dam was closed in 1932, converting this large body of water into a freshwater lake, the Dutch have been continually enclosing and reclaiming smaller bodies (polders). After dikes are built around a polder, the area is drained by pumping out the water. Drainage channels and, in many places, subsurface drains are installed so that the root zone of crops can be drained. After this, cropping is started as the last step in the reclamation process.

15 The development of land-clearing machinery and surface-drainage techniques has made it possible to clear and drain tropical lands for agricultural production. The first step is the removal of trees, brush, and other tropical growth. Outlet ditches are constructed,

followed by drains. In some cases subsurface drains are possible, but more often the soils and rainfall conditions combine to make this improvement impractical. Surface drains are installed on a uniform pattern and the land is smoothed or graded. Drainage systems on newly reclaimed tropical land require special attention while the soils are stabilizing, and some reconstruction is often needed after the soil stabilization is complete.

(1,149 words)

New Words

1. submain /ˈsʌbmeɪn/ n.   次干管（辅助干线）
2. outlet /ˈaʊtlet/ n.   排放孔，排水口
3. trapezoidal /ˌtræpɪˈzɔɪdəl/ adj.   梯形的
4. parallel /ˈpærəlel/ adj.   平行的
5. uniform /ˈjuːnɪfɔːm/ adj.   一致的
6. whereas /weərˈæz/ conj.   然而，鉴于
7. perpendicular /ˌpɜːpənˈdɪkjələ/ adj.   垂直的
8. emplace /ɪmˈpleɪs/ v.   安置就位
9. maintenance /ˈmeɪntənəns/ n.   维护
10. perforate /ˈpɜːfəreɪt/ v.   穿孔
11. rodent /ˈrəʊdənt/ n.   啮齿动物
12. reclaim /rɪˈkleɪm/ v.   开垦，改造
13. tropical /ˈtrɒpɪkəl/ adj.   热带的

Expressions

1. drainage system   排水系统
2. collecting ditch   集水沟
3. field ditch   毛渠，田沟
4. field drain   田间排水支沟
5. lateral conduit   分支管道

6	land grading	平整土地
7	drainage channel	排水沟
8	tile system	铺砌系统
9	bacterial action	细菌作用
10	precipitation pattern	降水年型
11	irrigated area	灌区，灌溉区
12	humid area	潮湿地区
13	root zone	根系层
14	excavation technology	挖掘技术
15	grade control	坡度控制
16	burrowing animal	洞穴动物

Note

Zuiderzee 须德海，原北海的海湾，在荷兰西北部。十三世纪时海水冲进内地，同原有湖沼汇合而成。

EXERCISES

I. Read each of the following paraphrases and then write the word it represents on the line provided.

1. a passage for escape or exit; a vent _____
2. being an equal distance apart everywhere _____
3. to put in place or position _____
4. the work of keeping something in proper condition; upkeep _____
5. of, occurring in, or characteristic of the tropics _____

II. Find the Chinese equivalents for the following expressions.

1. drainage system
2. field ditch
3. field drain
4. lateral conduit
5. drainage channel
6. bacterial action

7 precipitation pattern
8 irrigated area
9 humid area
10 root zone

III. **Discuss the following questions.**

1 How many types of drainage systems are there? What are they? What are their main components?
2 What could be the most common arrangements of field drains of a surface system?
3 What factors determine the arrangement of the laterals, the smallest component of the subsurface system?
4 How are the majority of underground drains built?
5 How can abandoned land resulted from salt accumulation be reclaimed?

IV. **Decide whether the following statements are true or false according to the text.**

_____ 1 The outlet is the point of disposal of water from the drainage system.
_____ 2 A constructed ditch tends to be rectangle in section.
_____ 3 Trees and shrubs can be planted on the banks of a river.
_____ 4 A surface system is the only choice for drainage improvement on soils that transmit water slowly.
_____ 5 The parallel system and random system both can lead to uniform drainage.
_____ 6 The computer models can help people select land grading.
_____ 7 The laterals can only be arranged in a uniform pattern.
_____ 8 Subsurface drainage systems are as significant in many irrigated areas as they are in humid areas.
_____ 9 Selected pesticides are often applied to keep woody growth and waterweed out of the channels.
_____ 10 It is not until the soil stabilization is complete that some reconstruction is required.

V. **Translate the following sentences into English.**

1 平行系统可以均匀地排水，而随机系统只能通过管道将水排至与之相连的低洼地区。(drain)
2 地下排水系统通常称之为铺砌系统，除排水口外，其他所有部分都位于地下。(surface of the ground)

3 传统来说，黏土或混凝土瓦管是建设地下排水沟的主要材料，但现在也会使用多种穿孔塑料管。(employ)

4 通常会在排水口上安装格栅，以防止啮齿动物和洞穴动物在此筑巢。(install)

5 几乎每年都需要对地表排水系统进行维护，以确保渠道的坡度和横截面以及平整区域的坡度。(maintenance)

VI. **Translate the passage into Chinese.**

The development of land-clearing machinery and surface-drainage techniques has made it possible to clear and drain tropical lands for agricultural production. The first step is the removal of trees, brush, and other tropical growth. Outlet ditches are constructed, followed by drains. In some cases subsurface drains are possible, but more often the soils and rainfall conditions combine to make this improvement impractical. Surface drains are installed on a uniform pattern and the land is smoothed or graded. Drainage systems on newly reclaimed tropical land require special attention while the soils are stabilizing, and some reconstruction is often needed after the soil stabilization is complete.

Unit 7
Dams

Lead-in

Dams are structures built across a stream, a river, or an estuary to retain water. They are built to increase the amount of water available for generating hydroelectric power, to reduce peak discharge of floodwater created by large storms or heavy snowmelt, or to increase the depth of water in a river in order to improve navigation and allow barges and ships to travel more easily. Dams can also provide a lake for recreational activities such as swimming, boating, and fishing. Many dams are built for more than one purpose.

Text A

Dams Are a Vital Part of the National Infrastructure

1. Water is one of our most precious resources; our lives depend on it. Throughout the history of humankind, people have built dams to maximize the use of this vital resource.

2. Dams provide a life-sustaining resource to people in all regions of the United States. They are an extremely important part of this nation's infrastructure—equal in importance to bridges, roads, airports, and other major elements of the infrastructure. They can serve several functions at once, including water supply for domestic, agricultural, industrial, and community use; flood control; recreation; and clean, renewable energy through hydropower.

3. As populations have grown and moved to arid or flood-prone locations, the need for dams has increased.

Benefits of Dams

4. Electrical generation: The U.S. is one of the largest producers of hydropower in the world, second only to Canada. Dams produce 8 to 12 percent of the nation's power needs.

5. Renewable, clean energy: Without hydropower, the U.S. would have to burn an additional 121 million tons of coal, 27 million barrels of oil, and 741 billion cubic feet of natural gas combined.

6. Flood control: Dams built with the assistance of the Natural Resources Conservation Service provide an estimated $1.7 billion in annual benefits in reduced flooding and erosion damage, recreation, water supplies, and wildlife habitat. Dams owned and operated by the Tennessee Valley Authority produce electricity and prevent an average of about $280 million in flood damage each year.

7. Water storage: Dams create reservoirs that supply water for a multitude of uses, including fire control, irrigation, recreation, domestic and industrial water supply, and more.

8. Irrigation: Ten percent of American cropland is irrigated using water stored behind dams.

9. Navigation: U.S. Army Corps of Engineers navigation projects in the U.S. serve 41

states, maintain 12,000 miles of channels, carry 15% of U.S. freight carried by inland waterways, operate 275 locks, and maintain 926 harbors.

10. Recreation: Dams provide prime recreational facilities throughout the U.S. Ten percent of the U.S. population visits at least one U.S. Army Corps of Engineers facility each year.

Dam Basics

11. The purpose of a dam is to impound (store) water, wastewater or liquid borne materials for any of several reasons, such as flood control, human water supply, irrigation, livestock water supply, energy generation, containment of mine tailings, recreation, or pollution control. Many dams fulfill a combination of the above functions.

12. Man-made dams may be classified according to the type of construction material used, the methods used in construction, the slope or cross-section of the dam, the way the dam resists the forces of the water pressure behind it, the means used for controlling seepage and, occasionally, according to the purpose of the dam.

13. The materials used for construction of dams include earth, rock, tailings from mining or milling, concrete, masonry, steel, timber, miscellaneous materials (such as plastic or rubber) and any combination of these materials.

Types of Dams

14. Embankment Dams: Embankment dams are the most common type of dam in use today. Materials used for embankment dams include natural soil or rock, or waste materials obtained from mining or milling operations. An embankment dam is termed an "earthfill" or "rockfill" dam depending on whether it is comprised of compacted earth or mostly compacted or dumped rock. The ability of an embankment dam to resist the reservoir water pressure is primarily a result of the mass weight, type and strength of the materials from which the dam is made.

15. Concrete Dams: Concrete dams may be categorized according to the designs used to resist the stress due to reservoir water pressure. Three common types of concrete dams are: gravity, buttress and arch.

16 Gravity: Concrete gravity dams are the most common form of concrete dams. The mass weight of concrete and friction resist the reservoir water pressure. Gravity dams are constructed of vertical blocks of concrete with flexible seals in the joints between the blocks.

17 Buttress: A buttress dam is a specific type of gravity dams in which the large mass of concrete is reduced, and the forces are diverted to the dam foundation through vertical or sloping buttresses.

18 Arch: Concrete arch dams are typically rather thin in cross-section. The reservoir water forces acting on an arch dam are carried laterally into the abutments. The shape of the arch may resemble a segment of a circle or an ellipse, and the arch may be curved in the vertical plane as well. Such dams are usually constructed of a series of thin vertical blocks that are keyed together; barriers to stop water from flowing are provided between blocks. Variations of arch dams include multi-arch dams in which more than one curved section is used, and arch-gravity dams which combine some features of the two types of dams.

Retaining Water and Seepage

19 Because the purpose of a dam is to retain water effectively and safely, the water retention ability of a dam is of prime importance. Water may pass from the reservoir to the downstream side of a dam by any of the following:

- Passing through the main spillway or outlet works.
- Passing over an auxiliary spillway.
- Overtopping the dam.
- Seepage through the abutments.
- Seepage under the dam.

20 Overtopping of an embankment dam is very undesirable because the embankment materials may be eroded away. Additionally, only a small number of concrete dams have been designed to be overtopped. Water normally passes through the main spillway or outlet works; it should pass over an auxiliary spillway only during periods of high reservoir levels and high water inflow. All embankment and most concrete dams have some seepage. However, it is important to control the seepage to prevent internal erosion and instability. Proper dam construction, and maintenance and monitoring of seepage provide this control.

Release of Water

21 Intentional release of water is confined to water releases through outlet works and spillways. A dam typically has a principal or mechanical spillway and a drawdown facility. Additionally, some dams are equipped with auxiliary spillways to manage extreme floods.

22 Outlet works: In addition to spillways that ensure that the reservoir does not overtop the dam, outlet works may be provided so that water can be drawn continuously, as needed, from the reservoir. They also provide a way to draw down the reservoir for repair or safety concerns. Water withdrawn may be discharged into the river below the dam, run through generators to provide hydroelectric power, or used for irrigation. Dam outlets usually consist of pipes, box culverts or tunnels with intake inverts near minimum reservoir level. Such outlets are provided with gates or valves to regulate the flow rate.

23 Spillways: The most common type of spillway is an ungated concrete chute. This chute may be located over the dam or through the abutment. To permit maximum use of storage volume, movable gates are sometimes installed above the crest to control discharge. Many smaller dams have a pipe and riser spillway, used to carry most flows, and a vegetated earth or rockcut spillway through an abutment to carry infrequent high flood flows. In dams such as those on the Mississippi River, flood discharges are of such magnitude that the spillway occupies the entire width of the dam and the overall structure appears as a succession of vertical piers supporting movable gates. High arch-type dams in rock canyons usually have downstream faces too steep for an overflow spillway. In Hoover Dam on the Colorado River, for example, a shaft spillway is used. In shaft spillways, a vertical shaft upstream from the dam drains water from the reservoir when the water level becomes high enough to enter the shaft or riser; the vertical shaft connects to a horizontal conduit through the dam or abutment into the river below.

(1,274 words)

New Words

1. infrastructure /ˈɪnfrəˌstrʌktʃə/ n. 基础设施
2. arid /ˈærɪd/ adj. 干旱的
3. habitat /ˈhæbɪtæt/ n. 栖息地
4. impound /ɪmˈpaʊnd/ v. 蓄水
5. containment /kənˈteɪnmənt/ n. 控制
6. tailing /ˈteɪlɪŋ/ n. 残渣
7. masonry /ˈmeɪsənri/ n. 砖石结构
8. miscellaneous /ˌmɪsəˈleɪniəs/ adj. 各种的
9. compacted /kəmˈpæktɪd/ adj. 夯实的，紧实的
10. friction /ˈfrɪkʃən/ n. 摩擦力
11. segment /ˈsegmənt/ n. 段
12. ellipse /ɪˈlɪps/ n. 椭圆
13. overtop /ˌəʊvəˈtɒp/ v. 漫溢
14. confine /kənˈfaɪn/ v. 限制，局限
15. chute /ʃuːt/ n. 斜槽，溜槽
16. crest /krest/ n. 顶部
17. magnitude /ˈmægnɪtjuːd/ n. 巨大
18. pier /pɪə/ n. 墩
19. shaft /ʃɑːft/ n. 竖井

Expressions

1. flood-prone 易发洪水的
2. embankment dam 土石坝
3. flexible seal 柔性密封
4. arch dam 拱坝
5. vertical plane 垂直面
6. outlet works 出水口工程
7. drawdown facility 排水设施
8. box culvert 箱涵
9. intake invert 进水口仰拱

10	flow rate	流量
11	shaft spillway	竖井式溢洪道
12	horizontal conduit	水平导管

Notes

1 **Natural Resources Conservation Service** 自然资源保护局（NRCS），之前的名称为土壤保护局（SCS），隶属美国农业部，负责向美国农场主、牧场主和森林所有者提供财务和技术协助，促进资源保护和农业运营。1994年比尔·克林顿（Bill Clinton）担任总统期间，该组织更名。其任务是通过与各州和地方机构的合作，改善和保护私人土地上的自然资源。虽然其主要任务为保护农业用地，但在土壤测量、分类和水质改善方面作出了许多技术贡献。

2 **U.S. Army Corps of Engineers** 美国陆军工兵部队，简称USACE，也称美国陆军工程兵团，是隶属于美国联邦政府和美国陆军的工兵单位，由美国陆军官兵和其他联邦文职人员组成，是世界最大的公共工程建设和建筑管理机构之一。

Exercises

I. Read each of the following paraphrases and then write the word it represents on the line provided.

1 extremely dry _____

2 the fundamental facilities serving a country, city, or area, such as transportation and communication systems, power supplies, and buildings _____

3 the natural environment in which a species or group of species lives _____

4 closely and firmly united or packed together _____

5 to keep within bounds _____

6 a usually curved structure forming the upper edge of an open space and supporting the weight above it, as in a bridge or doorway _____

7 surface resistance to relative motion, as of a body sliding or rolling _____

8 a geometrical figure that is a regular oval _____

9 a passageway through which surplus water escapes from a reservoir, lake, or the like _____

10 any device that shuts off, starts, regulates, or controls the flow of a fluid _____

II. Find the Chinese equivalents for the following expressions.

1. flood-prone
2. cross-section
3. embankment dam
4. flexible seal
5. arch dam
6. vertical plane
7. outlet works
8. drawdown facility
9. box culvert
10. intake invert

III. Discuss the following questions.

1. What functions can a dam serve?
2. What factors have increased the need for dams?
3. How can man-made dams be classified?
4. What materials can be used for construction of dams?
5. What types can dams be classified according to the materials used?
6. What are the common types of concrete dams?
7. Why is overtopping of an embankment dam undesirable?
8. Briefly describe the benefits of dams.

IV. Choose the best answer to each of the following questions or statements.

1. Dams can serve the following functions EXCEPT _____.

    A. bridges   B. flood control
    C. recreation   D. water supply

2. Which statement is true according to the text?

    A. More dams are needed partly because of the growth of population.
    B. More scenic spots are needed.
    C. More people have moved to flood-prone locations.
    D. More people have moved to dry areas.

3. Which of the following dams is the most common type in use today?

    A. Gravity dams.   B. Embankment dams.
    C. Buttress dams.   D. Arch dams.

4 Which of the following purposes is most important for a dam?
   A. Electricity generation.   B. Recreation.
   C. Retaining water.   D. Navigation.

5 The following measures are important for the control of the seepage of a dam EXCEPT _____.
   A. the construction of spillways   B. proper dam construction
   C. regular maintenance   D. monitoring of seepage

V. Translate the following sentences into Chinese.
   1 The ability of an embankment dam to resist the reservoir water pressure is primarily a result of the mass weight, type and strength of the materials from which the dam is made.
   2 Because the purpose of a dam is to retain water effectively and safely, the water retention ability of a dam is of prime importance.
   3 Concrete dams may be categorized according to the designs used to resist the stress due to reservoir water pressure.
   4 Materials used for embankment dams include natural soil or rock, or waste materials obtained from mining or milling operations.
   5 A dam typically has a principal or mechanical spillway and a drawdown facility.

VI. Translate the following sentences into English.
   1 水是我们赖以生存的最宝贵资源之一。(depend on)
   2 在人类历史上，人们一直在修建大坝以充分利用水资源。(make full use of)
   3 如果没有水电，人们不得不烧煤或油进行发电，这会对环境造成严重破坏。(do harm to)
   4 控制渗流对于防止大坝内部侵蚀和失稳非常重要。(internal erosion and instability)
   5 一些大坝配备了辅助溢洪道，以应对极端洪水。(equip with)

VII. Translate the passage into Chinese.
   Embankment dams are the most common type of dam in use today. Materials used for embankment dams include natural soil or rock, or waste materials obtained from mining or milling operations. An embankment dam is termed an "earthfill" or "rockfill" dam depending on whether it is comprised of compacted earth or mostly compacted or dumped rock. The ability of an embankment dam to resist the reservoir water pressure is primarily a result of the mass weight, type and strength of the materials from which the dam is made.

Text B

Dams: What They Are and What They Do

1. A dam is a structure built across a river or stream to hold back water. People have used different materials to build dams over the centuries. Ancient dam builders used natural materials such as rocks or clay. Modern-day dam builders often use concrete.

2. Man-made dams create artificial lakes called reservoirs. Reservoirs can be used to store water for farming, industry, and household use. They also can be used for fishing, boating, and other leisure activities. People have used dams for many centuries to help prevent flooding.

3. Dams have two main functions. The first is to store water to compensate for fluctuations in river flow or in demand for water and energy. The second to raise the level of the water upstream to enable water to be diverted into a canal or to increase "hydraulic head"—the difference in height between the surface of a reservoir and the river downstream. The creation of storage and head allows dams to generate electricity (hydropower provides nearly a fifth of the world's electricity); to supply water for agriculture, industries and households; to control flooding; and to assist river navigation by providing regular flows and drowning rapids. Other reasons for building large dams include reservoir fisheries and leisure activities such as boating.

4. Hydropower generation capacity is a function of the amount of flow and hydraulic head. Although the head is usually related to the height of the dam, a low dam can have a high head if the powerhouse with its turbines and generators is located some distance downstream of the dam. Pipes known as "penstocks" direct water to the turbines. Once the water has spun a turbine, it flows into the "tailwater" below the dam through a "tailrace" pipe.

5. One advantage of hydro over other forms of electricity generation is that reservoirs can store water during times of low demand and then quickly start generating during the peak hours of electricity use. Thermal power plants take much longer to start up from cold than hydro plants. Hydro's suitability for generating valuable "peaking" power has in recent

years encouraged a boom in what are known as pumped-storage plants. These involve two, normally relatively small, reservoirs, one above the other. During peak hours, the water from the upper reservoir falls through turbines into the lower one, generating electricity. The water is then pumped back uphill again using cheap off-peak electricity.

6. Weirs and barrages are different types of "run-of-river" dams, this means that while they raise the water level upstream they create only a small reservoir ("head pond") and cannot effectively regulate downstream flows. A weir is normally a low wall of stone, concrete or wicker. A barrage can be a huge structure ten or twenty meters high extending for hundreds of meters across the bottom reaches of a wide river. The electricity generation of a "run-of-river" hydropower dam is proportional to the flow of the river at any one time.

7. While they tend to have less damaging consequences than storage dams, run-of-river dams are far from environmentally benign, and the distinction between a "run-of-river" and a "storage" dam is not always clear. Dam proponents have in some cases sought to downplay the impact of planned dams by claiming that they will be run-of-river. Thailand's Pak Mun Dam, for example, is repeatedly described by officials as a run-of-river project yet for much of the time the dam's gates remain closed and it operates as a storage dam. Despite years of protestations from its builders and funders that it would have minimal impacts on the river, Pak Mun managed within a couple of years to destroy one of the country's richest freshwater fisheries.

8. Just as every river and watershed is unique, so is every dam site and every dam. There are, however, three main types of dam design—embankment, gravity and arch—selected mainly according to dam-site topography and geology. Earth and rock embankments, which are usually the cheapest to build, make up more than 80 percent of all large dams. Embankments are generally built across broad valleys near sites where the large amounts of construction material they need can be quarried. Large embankment dams are the most massive structures humanity has ever erected. The most voluminous dam in the world, Tarbela in Pakistan, contains 106 million cubic meters of earth and rock, more than 40 times the volume of the Great Pyramid.

9. Gravity dams are basically thick, straight walls of concrete built across relatively narrow valleys with firm bedrock. Arch structures, also made from concrete, are limited to narrow canyons with strong rock walls and make up only around four percent of large dams. An

arch dam is in form like a normal architectural arch pushed onto its back, with its curved top facing upstream and its feet braced against the sides of its canyon. The inherent strength of the shape enables the thin wall of an arch dam to hold back a reservoir with only a fraction of the concrete needed for a gravity dam of similar height.

10 A dam contains a number of structural features other than the main wall itself. Spillways are used to discharge water when the reservoir threatens to become dangerously high. Dams built across broad plains may include long lengths of ancillary dams and dykes. The five reservoirs of Phase I of the La Grande hydropower scheme in northern Quebec, for example, are impounded by eleven dams and more than 200 accompanying dykes stretching for a total length of 124 kilometers.

(910 words)

New Words

1. fluctuation /ˌflʌktʃuˈeɪʃən/ n.   波动
2. fishery /ˈfɪʃəri/ n.   渔业
3. penstock /ˈpenstɒk/ n.   压力水管
4. tailwater /ˈteɪlˌwɔːtə/ n.   下游水（尾水；废水）
5. tailrace /ˈteɪlreɪs/ n.   尾水渠
6. barrage /ˈbærɑːʒ/ n.   拦河坝
7. wicker /ˈwɪkə/ n.   枝条
8. benign /bɪˈnaɪn/ adj.   良性的
9. proponent /prəˈpəʊnənt/ n.   支持者
10. downplay /ˌdaʊnˈpleɪ/ v.   低估
11. topography /təˈpɒɡrəfi/ n.   地势，地形
12. geology /dʒiˈɒlədʒi/ n.   （某地区的）地质情况
13. quarry /ˈkwɒri/ v.   挖出
14. voluminous /vəˈluːmɪnəs/ adj.   大量的
15. brace /breɪs/ v.   支撑
16. ancillary /ænˈsɪləri/ adj.   辅助的
17. dyke /daɪk/ n.   堤坝

Expressions

1	drowning rapids	急流
2	thermal power plant	火力发电厂
3	pumped-storage plant	抽水蓄能电站
4	off-peak	非峰期

Notes

1 **Pak Mun Dam** 巴蒙大坝，位于湄公河支流门河（Mun River）的泰国巴蒙大坝1994年完工。该坝通过建造鱼梯，试图减轻鱼类无法自然迁徙而造成的环境冲击，结果证明是一无是处。水库上游的可捕鱼量持续锐减，环境资源遭到摧毁性破坏，危及成千上万泰国当地民众生计。泰国巴蒙大坝的建造，导致渔业资源枯竭的危害，让泰国新的大坝计划尽量设计一些帮助鱼类迁徙的友善设施，但无法获得预期效果。

2 **Tarbela Dam** 塔贝拉坝，位于巴基斯坦、印度河干流。距离首都伊斯兰堡约2.5小时的车程，拉瓦尔品第西北约64公里。工程具有灌溉、发电、防洪等效益，是印度河西水东调的关键工程。

Exercises

I. Read each of the following paraphrases and then write the word it represents on the line provided.

1 the quality of being unsteady and subject to changes _____

2 an artificial obstruction, such as a dam or irrigation channel, built in a watercourse to increase its depth or to divert its flow _____

3 having little or no detrimental effect; harmless _____

4 one who argues in support of something; an advocate _____

5 to play down; make little of _____

6 business or industry of fishing _____

7 the surface features of a place or region _____

8 the structure of a specific region of the Earth, including its rocks, soils, mountains, fossils, and other features _____

9 something that is subordinate or accessory to something else _____

10 an embankment or wall built to confine a river to a particular course _____

II. **Discuss the following questions.**

1. What can reservoirs be used for?
2. What are the two main functions of dams?
3. What does hydropower generation capacity depend on?
4. What are pumped-storage plants characterized by?
5. What are run-of-river dams characterized by?
6. How many main types of dam design are there? What are they?

III. **Decide whether the following statements are true or false according to the text.**

_____ 1 Materials such as rocks, clay, concrete and wicker are often used to construct a dam.
_____ 2 The water head depends on the height of a dam, so a low dam is unlikely to have a high head.
_____ 3 Thermal power plants take much more time to start up from cold than hydro plants.
_____ 4 The electricity produced by a "run-of-river" hydropower dam depends on the flow of the river at any one time.
_____ 5 Run-of-river dams are as unfriendly as storage dams to environment.
_____ 6 Pak Mun Dam destroyed one of Thailand's richest freshwater fisheries.
_____ 7 Most of large dams are embankment dams owing to cheap materials.
_____ 8 A gravity dam usually needs less concrete than an arch dam of similar height.
_____ 9 Small reservoirs can be created by barrages which belong to storage dams.
_____ 10 Generally, a storage dam does greater harm to environment than a run-of-river dam.

IV. **Translate the following sentences into English.**

1. 大型土石坝是人类有史以来建造的最庞大的建筑。(erect)
2. 水库可蓄水，用于农业、工业、家庭、渔业、划船和其他休闲活动。(store water)
3. 水力发电量取决于流量和水头。(capacity)
4. 抽水蓄能电站可以在用电需求低时段蓄水，然后在用电高峰时段快速发电。(generate)
5. 溢洪道用于在水库可能达到危险高度时排水。(discharge)

V. **Translate the passage into Chinese.**

Dams have two main functions. The first is to store water to compensate for fluctuations in river flow or in demand for water and energy. The second to raise the level of the water upstream to enable water to be diverted into a canal or to increase "hydraulic head"—

the difference in height between the surface of a reservoir and the river downstream. The creation of storage and head allows dams to generate electricity (hydropower provides nearly a fifth of the world's electricity); to supply water for agriculture, industries and households; to control flooding; and to assist river navigation by providing regular flows and drowning rapids. Other reasons for building large dams include reservoir fisheries and leisure activities such as boating.

Unit 8

Hydraulic Turbines

Lead-in

A water turbine is a rotary machine that converts kinetic energy and potential energy of water into mechanical work. Water turbines were developed in the 19th century and were widely used for industrial power prior to electrical grids. Now they are mostly used for electric power generation. Water turbines are mostly found in dams to generate electric power from water potential energy. There are two main types of hydro turbines: impulse and reaction. The type of hydropower turbine selected for a project is based on the height of standing water—referred to as "head"—and the flow, or volume of water, at the site.

Text A

Hydro Turbine

1. Hydro turbines are devices used in hydroelectric generation plants that transfer the energy from moving water to a rotating shaft to generate electricity. These turbines rotate or spin as a response to water being introduced to their blades. These turbines are essential in the area of hydropower—the process of generating power from water.

2. Generally, the construction of turbines is the same. A row of blades is fitted to some rotating shaft or plate. Water is then passed through the turbine over the blades, causing the inner shaft to rotate. This rotational motion is then transferred to a generator where electricity is generated. There are a variety of different types of turbines that are best used in different situations. Each type of turbine is created to provide maximum output for the situation it is used in.

3. The type of turbine selected for any given hydropower project is based on the height and speed of the incoming water—known as the hydraulic head—and the volume of water that flows known as the hydroelectric discharge. Efficiency and cost are also factors to be considered.

Turbine Flow

4. Hydro turbines can be classified based on how water flows through the turbine itself. When passing through a turbine, water can take a variety of different paths. This leads to three categories of water flow through turbines:

- Axial flow: Water flows through the turbine parallel to the axis of rotation.

- Radial flow: Water flows through the turbine perpendicular to the axis of rotation.

- Mixed flow: Water flows through in a combination of both radial and axial flows. For example, in a Francis turbine water flows in radially but exits axially. Most hydro turbines tend to have mixed flows.

Pressure Change

5   Another criteria used to classify turbines is whether or not the liquid pressure changes when flowing through a turbine. There are two types of turbines that emerge from this classification, explained below.

Impulse Turbine

6   In impulse turbines, the pressure of the liquid doesn't change in the turbine itself. Instead, pressure changes occur only in the nozzles that direct water flow to the turbine, while the turbine itself operates at atmospheric pressure. These turbines are composed of a jet nozzle or series of nozzles that direct water to the blades of a turbine. Multiple nozzles are usually used where a large wheel isn't feasible. When the water strikes the blades (designed specially to reduce drag), it changes velocity. This leads to a change in momentum, exerting a force on the turbine blades. These turbines rely on the ability to take all kinetic energy from the water to have high efficiencies. Unlike reaction turbines, impulse turbines do not need to be submerged. Types of impulse turbines include Pelton turbines, Turgo turbines, and Crossflow turbines.

7   A Pelton turbine or Pelton wheel is a type of hydro turbine (specifically an impulse turbine) used frequently in hydroelectric plants. These turbines are generally used for sites with heads greater than 300 meters. This type of turbine was created during the gold rush in 1880 by Lester Pelton. The water in a Pelton turbine is moving quickly and the turbine extracts energy from the water by slowing the water down, which makes this an impulse turbine.

8   A Turgo turbine is a type of hydro turbine (specifically an impulse turbine), developed in the 1920s, which is a variant of the Pelton turbine. The main difference between a Pelton turbine and a Turgo turbine is that Turgo turbines use single cups instead of double cups on the wheel, and these cups are more shallow. The water in a Turgo turbine is moving quickly and the turbine extracts energy from the water by slowing the water down, which makes this an impulse turbine.

9   Crossflow turbines are a type of hydro turbines (specifically an impulse turbine) that tends to be used in smaller hydroelectric sites with power outputs between 5-100kW. These turbines are useful for a large range of hydraulic heads, from only 1.75 meters to 200 meters, although usually crossflow turbines are chosen for heads below 40 meters.

Crossflow turbines get energy from water by reducing the velocity, the pressure stays the same, which is why they're impulse turbines.

Reaction Turbine

10 In reaction turbines, the pressure of the water changes as it moves through the blades of the turbine. Unlike in an impulse turbine, the reaction turbine directly uses water pressure to move blades instead of relying on a conversion of water pressure to kinetic energy. This means that reaction turbines need to be submersed in water. Additionally, components of these turbines must be able to handle high pressure levels inside the turbine. Here, the fluid velocity and reduction in water pressure causes a reaction on the turbine blades, moving them. Types of reaction turbines include Kaplan turbines and Francis turbines.

11 A Kaplan turbine is a type of propeller hydro turbine (specifically a reaction turbine) used in hydroelectric plants. Water flows both in and out of Kaplan turbines along its rotational axis (axial flow). What makes Kaplan turbines special is the blades can change their angle on demand to maintain maximum efficiency for different flow rates of water. Water flowing through a Kaplan turbine loses pressure, and this means that a Kaplan turbine is a reaction turbine (similar to a Francis turbine).

12 A Francis turbine is a type of reaction turbines used most frequently in medium- or large-scale hydroelectric plants. These turbines can be used for heads as low as 2 meters and as high as 300 meters. Additionally, these turbines are beneficial as they work equally well when positioned horizontally as they do when they are oriented vertically. Francis turbines are the most frequently used turbines for hydropower plants. The water going through a Francis turbine loses pressure, but stays at more or less the same speed, so it would be considered a reaction turbine.

13 Water entering these turbines radially means that it enters the turbine perpendicular to the rotational axis. Once entering the turbine, the water always flows inwards, towards the center. Once the water has flown through the turbine, it exits axially—parallel to the rotational axis. Francis turbines were the first hydraulic turbines that had a radial inflow, designed by American scientist James Francis.

(1,035 words)

Unit 8　Hydraulic Turbines

New Words

1. rotate /rəʊˈteɪt/ v.　（使）旋转
2. spin /spɪn/ v.　（使）旋转
3. axial /ˈæksiəl/ adj.　轴的
4. radial /ˈreɪdiəl/ adj.　放射状的
5. nozzle /ˈnɒzəl/ n.　喷嘴
6. velocity /vəˈlɒsəti/ n.　速度
7. momentum /məʊˈmentəm/ n.　动量
8. submerge /səbˈmɜːdʒ/ v.　（使）浸没
9. variant /ˈveəriənt/ n.　变体
10. propeller /prəˈpelə/ n.　螺旋桨

Expressions

1. kinetic energy　动能
2. generation plants　发电站
3. hydraulic head　水头
4. hydroelectric discharge　水力发电流量
5. axial flow　轴向流动
6. radial flow　径向流动
7. mixed flow　混合流动
8. impulse turbine　冲击式水轮机
9. atmospheric pressure　气压
10. jet nozzle　射流喷嘴
11. reaction turbine　反击式水轮机
12. crossflow turbine　横流式水轮机
13. power output　输出功率

Exercises

I. Read each of the following paraphrases and then write the word it represents on the line provided.

1. a projecting part with an opening, as at the end of a hose, for regulating and directing a flow of fluid _____
2. a revolving rod that transmits power or motion _____
3. to turn around on an axis or a center _____
4. to place under the surface of a liquid, especially water _____
5. a device having blades radiating from a central hub that is rotated to produce thrust to propel a ship, aircraft, etc. _____

II. Find the Chinese equivalents for the following expressions.

1. hydraulic head
2. axial flow
3. radial flow
4. mixed flow
5. impulse turbine
6. atmospheric pressure
7. jet nozzle
8. kinetic energy
9. crossflow turbine
10. reaction turbine

III. Discuss the following questions.

1. How is a turbine constructed?
2. How can turbines be classified?
3. In impulse turbines, where do water pressure changes occur?
4. Which kind of turbines needs to be submerged in water?
5. Which type is a Pelton turbine classified to? What sites is it suitable for?
6. Describe the main types and characteristics of hydraulic turbines.

IV. Choose the best answer to each of the following questions or statements.

1. The type of turbine selected for any given hydropower project is based on the following factors EXCEPT _____.
   A. hydraulic head　　　　　　B. evaporation
   C. cost　　　　　　　　　　　D. efficiency

2. Types of impulse turbines include the following EXCEPT _____.
   A. Kaplan turbines　　　　　　B. Pelton turbines
   C. Turgo turbines　　　　　　 D. Crossflow turbines

3. If a small hydropower plant with low water head is to be constructed, _____ are most suitable for the project.
   A. Kaplan turbines　　　　　　B. Pelton turbines
   C. Turgo turbines　　　　　　 D. Crossflow turbines

4. A Kaplan turbine works in a similar way with a _____.
   A. Pelton turbine　　　　　　 B. Turgo turbine
   C. Francis turbine　　　　　　D. Crossflow turbine

5. Which statement is NOT true about a reaction turbine?
   A. The pressure of the water changes as it moves through the blades of the turbine.
   B. A reaction turbine directly uses water pressure to move blades.
   C. A reaction turbine needs to be submersed in water.
   D. A reaction turbine is composed of a jet nozzle or series of nozzles that direct water to its blades.

V. Translate the following sentences into Chinese.

1. Hydro turbines are devices used in hydroelectric generation plants that transfer the energy from moving water to a rotating shaft to generate electricity.

2. Unlike in an impulse turbine, the reaction turbine directly uses water pressure to move blades instead of relying on a conversion of water pressure to kinetic energy.

3. What makes Kaplan turbines special is the blades can change their angle on demand to maintain maximum efficiency for different flow rates of water.

4. Additionally, these turbines are beneficial as they work equally well when positioned horizontally as they do when they are oriented vertically.

5. Once the water has flown through the turbine, it exits axially—parallel to the rotational axis.

VI. Translate the following sentences into English.
1 水轮机的选择取决于使用时能否提供最大输出。(maximum output)
2 水轮机是由其叶片传动的，叶片是由水流驱动的。(rotate)
3 水轮机类型的选择依据的是水头、流量、效率和成本因素。(be based on)
4 水轮机可以根据水流通过水轮机本身的方式进行分类。(be classified by)
5 与冲击式水轮机不同，反击式水轮机需要浸没在水中。(submerged)

VII. Translate the passage into Chinese.

Generally, the construction of turbines is the same. A row of blades is fitted to some rotating shaft or plate. Water is then passed through the turbine over the blades, causing the inner shaft to rotate. This rotational motion is then transferred to a generator where electricity is generated. There are a variety of different types of turbines that are best used in different situations. Each type of turbine is created to provide maximum output for the situation it is used in.

Text B

Factors Affecting the Selection of Hydraulic Turbines

1 Hydraulic turbines or water turbines are important hydraulic machines of the hydropower plant that convert the hydraulic energy into mechanical energy. There are different types of turbines such as a Pelton turbine, Kaplan turbine, Francis turbine, bulb turbine etc.

2 Hydraulic turbines may be grouped in two general classes: the impulse type which utilizes the kinetic energy of a high-velocity jet which acts upon only a small part of the circumference at any instant, and the reaction type which develops power from the combined action of pressure and velocity of the water that completely fills the runner and water passages. The reaction group is divided into two general types: the Francis, sometimes called the reaction type, and the propeller type. The propeller class is also further subdivided into the fixed blade or propeller type, and the adjustable-blade type of which the Kaplan is representative.

3 Power may be developed from water by three fundamental processes: by action of its weight, of its pressure, or of its velocity; or by a combination of any or all three. In modern practice the Pelton or impulse wheel is the only type which obtains power by a single process, the action of one or more high-velocity jets. This type of wheel is usually found in high-head developments.

4 The problem confronting engineers is to determine the type and setting that will result in the most economical development under the given existing conditions. Each type of turbine has its limitations in application. Following are the factors affecting the selection of hydraulic turbines:

(1) head

(2) specific speed

(3) rotational speed

(4) efficiency of the turbine

(5) cavitation

(6) disposition of the turbine shaft

(7) part-load operation

(1) Head

5 The hydraulic turbine selected should be decided based on the head, and for different head ranges different turbines are suitable. Heads are divided into five types and suitable turbines for them are given in the table below.

Head	Head Range in Meters	Suitable Turbine	Notes
Very low head	3 – 10	Bulb turbine	Kaplan turbines are also suitable but uneconomical for very low heads.
Low head	10 – 60	Kaplan turbine	Propeller turbines are also suitable up to 15m head but there should not be load variations.

(Continued)

Head	Head Range in Meters	Suitable Turbine	Notes
Medium head	60 – 150	Francis turbine	/
High head	150 – 350	Pelton or Francis turbine	One of them is decided based on the specific speed.
Very high head	>350	Pelton turbine	/

(2) Specific Speed

6　The specific speed is high for a turbine which works under high head and low flow rate. Similarly, the specific speed is low for a turbine which works under low head and high flow rate. When the specific speed of the selected turbine is not up to the required range, then the rotational speed of turbine reduces, and thus the capacity of plant reduces. However, medium-specific speed turbines such as Francis turbines can generate high rotational speed.

Type of Turbine	Specific Speed Range
Pelton turbine	10 – 70
Francis turbine	80 – 400
Kaplan turbine	300 – 1000
Bulb turbine	1000 – 1200

(3) Rotational Speed

7　The rotational speed of the turbine depends upon the specific speed of the turbine, frequency and number of pair of poles in the electric generator. So, the specific speed of the selected turbine should produce the same amount of rotation speed of the generator.

(4) Efficiency of Turbine

8　The efficiency of the turbine selected should be as high as possible and it is considered for various working conditions of turbines. Different turbines in decreasing order of their overall efficiency are as follows:

Francis turbine > Kaplan turbine > Pelton turbine

(5) Cavitation

9　Cavitation mainly occurs in the case of reaction type turbines. So, while selecting the turbine, the cavitation factor should be determined for that particular turbine to check whether it is in a safe zone or not. Cavitation factors depend upon the specific speed of the turbine. It is important, especially in the case of Kaplan, propeller and bulb type turbines.

(6) Disposition of Turbine Shaft

10 Based on the previous experiences, it is recommended that the horizontal shaft arrangement is best suitable for large size impulse turbines such as Pelton turbine etc. Similarly in case of large size reaction turbines such as the Kaplan turbine, the Deriaz turbine etc., the vertical shaft arrangement is recommended.

(7) Part-load Operation

11 In general, the efficiency of the turbine is maximum when it is running with a designed load condition. When the part-load or overload condition arises, the efficiency reduces, which is uneconomical. In that case, Deriaz turbine or Kaplan turbine is recommended.

(753 words)

New Words

1	convert /kənˈvɜːt/ v.	转化
2	utilize /ˈjuːtəlaɪz/ v.	利用
3	subdivide /ˌsʌbdɪˈvaɪd/ v.	细分
4	representative /ˌreprɪˈzentətɪv/ adj.	典型的
5	confront /kənˈfrʌnt/ v.	面临
6	setting /ˈsetɪŋ/ n.	挡位
7	pole /pəʊl/ n.	极点
8	cavitation /ˌkævɪˈteɪʃən/ n.	汽蚀
9	disposition /ˌdɪspəˈzɪʃən/ n.	排列
10	vertical /ˈvɜːtɪkəl/ adj.	直立的
11	overload /ˌəʊvəˈləʊd/ v.	使超负荷

Expressions

1	bulb turbine	灯泡贯流式水轮机
2	high-velocity jet	高速射流
3	water passage	水道
4	adjustable-blade	可调叶片
5	high-head	高水头
6	specific speed	比转速
7	rotational speed	转速

Exercises

I. Read each of the following paraphrases and then write the word it represents on the line provided.

1. to change (something) into another form, substance, state, or product _____
2. to put to use, especially to make profitable or effective use of _____
3. to divide a part or parts of something into smaller parts _____
4. to face with _____
5. upright in position or posture _____

II. Find the Chinese equivalents for the following expressions.

1. high-velocity jet
2. water passage
3. adjustable-blade
4. high-head
5. specific speed

III. Discuss the following questions.

1. How many types can hydraulic turbines be classified into?
2. What factors should be considered when we select hydraulic turbines?
3. How many types can reaction turbines be divided into? What are they?
4. What types can propeller turbines be subdivided into? What are they?
5. By what processes can power be developed from water?

IV. Decide whether the following statements are true or false according to the text.

_____ 1 Hydraulic turbines convert water energy into electricity.

_____ 2 Water head is one of the key factors affecting the selection of hydraulic turbines.

_____ 3 Efficiency is the only factor to be considered when we select water turbines.

_____ 4 Cavitation usually occurs in the case of impulse turbines.

_____ 5 When the part-load or overload condition arises the efficiency of water turbines reduces.

V. Translate the following sentences into English.
1. 水轮机可以把水能转换成机械能。(convert)
2. 水轮机的选择需要考虑多种因素。(consider)
3. 所选水轮机的效率应尽量高，并考虑到水轮机的各种工况。(working conditions)
4. 根据以往的经验，建议大型冲击式水轮机的旋转轴最好采用横轴。(horizontal shaft arrangement)
5. 如果是大型反击式水轮机，建议采取立轴。(vertical shaft arrangement)

VI. Translate the passage into Chinese.

Hydraulic turbines may be grouped in two general classes: the impulse type which utilizes the kinetic energy of a high-velocity jet which acts upon only a small part of the circumference at any instant, and the reaction type which develops power from the combined action of pressure and velocity of the water that completely fills the runner and water passages. The reaction group is divided into two general types: the Francis, sometimes called the reaction type, and the propeller type. The propeller class is also further subdivided into the fixed blade or propeller type, and the adjustable-blade type of which the Kaplan is representative.

Unit 9
Hydroelectric Power

Lead-in

Hydropower is a form of energy that harnesses the power of water in motion—such as water flowing over a waterfall—to generate electricity. There are several types of hydroelectric facilities; they are all powered by the kinetic energy of flowing water as it moves downstream. Turbines and generators convert the energy into electricity, which is then fed into the electrical grid to be used in homes, businesses, and industry.

Text A

Hydropower Today

1. Hydropower, also known as hydroelectric power, is a reliable, domestic, emission-free resource that is renewable through the hydrologic cycle and harnesses the natural energy of flowing water to provide clean, fast and flexible electricity generation.

2. Because hydropower generation begins the minute water starts to fall through the turbines, it is capable of rapid response to peak demands and emergency needs, contributing to the stability of a nation's electricity grid and energy security. Hydropower is also one of the most economic energy resources and is not subject to market fluctuations or embargos, which helps support the nation's energy independence—and it can provide that support for years to come. The average lifespan of a hydropower facility is 100 years. By upgrading and increasing the efficiencies and capacities of existing facilities, hydropower can continue to support the nation's growing energy needs. Hydropower also has more non-power benefits than any other generation sources, including water supply, flood control, navigation, irrigation, and recreation. While there are many advantages to hydroelectric production, the industry also faces unique environmental challenges. Potential environmental impacts include changes in aquatic and stream side habitats; alteration of landscapes through the formation of reservoirs; effects on water quality and quantity; interruption of migratory patterns for fish such as salmon, steelhead, American shad, and sturgeon; and injury or death to fish passing through the turbines. The challenge facing hydropower researchers today is how to take advantage of one of the nation's most plentiful renewable resources to produce the electricity we need without endangering the aquatic species and habitats upon which the health of the environment and industries such as fishing and river tourism depend. Unless they find a way to meet this challenge, researchers and industry members feel it is unlikely that much additional hydropower will be added to our generation mix through undeveloped resources.

How Hydropower Works

3. Hydropower plants capture the energy of falling water to generate electricity. A turbine

converts the kinetic energy of falling water into mechanical energy. Then a generator converts the mechanical energy from the turbine into electrical energy.

Parts of a Hydroelectric Plant

4 Most conventional hydroelectric plants include four major components:

(1) Dam. A dam raises the water level of the river to create falling water. Also it controls the flow of water. The reservoir that is formed is, in effect, stored energy.

(2) Turbine. The force of falling water pushing against the turbine's blades causes the turbine to spin. A water turbine is much like a windmill, except the energy is provided by falling water instead of wind. The turbine converts the kinetic energy of falling water into mechanical energy.

(3) Generator. A generator is connected to the turbine by shafts and possibly gears so when the turbine spins it causes the generator to spin also. It converts the mechanical energy from the turbine into electric energy. Generators in hydropower plants work just like the generators in other types of power plants.

(4) Transmission lines. They conduct electricity from the hydropower plant to homes and business.

Types of Hydropower Plants

5 Many dams were built for other purposes and hydropower was added later. Some dams are for recreation, stock/farm ponds, flood control, water supply, and irrigation. Hydropower plants range in size from small systems for a home or village to large projects producing electricity for utilities. There are four types of hydropower facilities: impoundment, diversion, run-of-river, and pumped storage. Some hydropower plants use dams and some do not.

Diversion

6 A diversion facility channels a portion of a river through a canal or penstock. It may not require the use of a dam.

Run-of-River

7 A run-of-river project uses water within the natural flow range of the river, requiring little or no impoundment.

Pumped Storage

8 When the demand for electricity is low, a pumped storage facility stores energy by pumping water from a lower reservoir to an upper reservoir. During periods of high electrical demand, the water is released back to the lower reservoir to generate electricity.

Impoundment

9 The most common type of hydroelectric power plant is an impoundment facility. An impoundment facility, typically a large hydropower system, uses a dam to store river water in a reservoir. Water released from the reservoir flows through a turbine, spinning it, which in turn activates a generator to produce electricity. The water may be released either to meet changing electricity needs or to maintain a constant reservoir level.

10 An impoundment hydropower plant uses a dam to store water in a reservoir.

Sizes of Hydroelectric Power Plants

11 Facilities range in size from large power plants that supply many consumers with electricity to small and micro plants that individuals operate for their own energy needs or to sell power to utilities.

Large Hydropower

12 Although definitions vary, large hydropower refers to facilities that have a capacity of more than 30 megawatts (MW).

Small Hydropower

13 Although definitions vary, small hydropower means projects that generate 10 MW or less of power.

Micro Hydropower

14 A micro hydropower plant has a capacity of up to 100 kilowatts. A small or micro-hydroelectric power system can produce enough electricity for a home, farm, ranch, or village.

Transmitting Power

15 Once the electricity is produced, it must be delivered to where it is needed—our homes, schools, offices, factories, etc. Dams are often in remote locations and power must be transmitted over some distance to its users.

16 Vast networks of transmission lines and facilities are used to bring electricity to us

in a form we can use. All the electricity made at a powerplant comes first through transformers which raise the voltage so it can travel long distances through powerlines. At local substations, transformers reduce the voltage so electricity can be divided up and directed throughout an area. Transformers on poles (or buried underground, in some neighborhoods) further reduce the electric power to the right voltage for appliances and use in the home. When electricity gets to our homes, we buy it by the kilowatt-hour, and a meter measures how much we use.

17 While hydroelectric powerplants are one source of electricity, other sources include powerplants that burn fossil fuels or split atoms to create steam which in turn is used to generate power. Gas turbine, solar, geothermal, and wind-powered systems are other sources. All these powerplants may use the same system of transmission lines and stations in an area to bring power to you. By use of this "power grid", electricity can be interchanged among several utility systems to meet varying demands. So the electricity lighting your reading lamp now may be from a hydroelectric powerplant, a wind generator, a nuclear facility, or a coal, gas, or oil-fired powerplant… or a combination of these.

(1,103 words)

New Words

1. harness /ˈhɑːnɪs/ v. 利用
2. kinetic /kɪˈnetɪk/ adj. 运动的
3. embargo /ɪmˈbɑːgəʊ/ n. 禁运
4. windmill /ˈwɪndˌmɪl/ n. 风车
5. stock /stɒk/ n. 储备
6. diversion /daɪˈvɜːʃən/ n. 引水
7. megawatt /ˈmegəwɒt/ n. 【电】兆瓦（特）
8. transformer /trænsˈfɔːmə/ n. 变压器
9. geothermal /ˌdʒiːəʊˈθɜːməl/ adj. 地热的

Expressions

1 hydrologic cycle　　　　　　　水循环
2 hydropower plant　　　　　　 水力发电厂
3 pumped storage　　　　　　　抽水蓄能
4 power grid　　　　　　　　　电网

Notes

1. **Diversion Hydropower Plant** 引水式水电厂，是水电开发的基本形式之一。这类水电厂宜建在河道坡降较陡的河段或大河湾处，在河段上游筑坝引水，用引水渠道、隧洞、压力水管等将水引到河段下游，用以集中水头发电。这类水电厂大都为高水头水电厂。跨流域引水发电的水电厂必然是引水式水电厂。

2. **Run-of-River Hydropower Plant** 径流式水电厂，是在河道中拦河筑低坝或闸，基本不调节径流，靠天然径流发电的水电厂。当来水流量大于水轮机过水能力时，水电厂满出力运行，多余水量经泄水建筑物直接泄向下游，称为弃水；当来水流量小于水轮机过水能力时，则有部分发电机组容量能被利用。水电厂的这种运行方式称为径流发电。

3. **Pumped Storage Hydropower Plant** 抽水蓄能电厂，建有上下两座水库，其间用压力隧洞或压力水管相连接。它是利用电力系统低谷负荷时的剩余电力从下库抽水蓄存到上库，在高峰负荷时从上库放水至下库发电的水电厂。

4. **Impoundment Hydropower Plant** 蓄水式水电厂，或称坝式水电厂，是水电开发的基本方式之一。蓄水式水电厂是由河道上的挡水建筑物壅高水位而集中水头的水电厂。当水头不高且河道较宽时，用厂房作为挡水建筑物的一部分，这类水电厂又称河床式水电厂。

Exercises

I. Read each of the following paraphrases and then write the word it represents on the line provided.

  1 a prohibition by a government on certain or all trade with a foreign nation _____
  2 a machine that runs on the energy generated by a wheel of adjustable blades or slats rotated by the wind _____
  3 a pipe or conduit used to carry water to a water wheel or turbine _____
  4 a device that transfers an alternating current from one circuit to one or more other circuits, usually with an increase or decrease of voltage _____
  5 of or relating to the internal heat of the Earth _____

II. Find the Chinese equivalents for the following expressions.

1 hydrologic cycle
2 turbine
3 pumped storage
4 power grid
5 hydropower plant

III. Discuss the following questions.

1 What are the advantages of hydropower?
2 What environmental challenges does hydroelectric production face?
3 What parts do most conventional hydroelectric plants include?
4 What types of hydropower plants are there?
5 How is electricity transmitted to our homes?
6 Describe how hydropower is generated.

IV. Choose the best answer to each of the following questions or statements.

1 The average lifespan of a hydropower facility is _____ years.
   A. 50   B. 100
   C. 150   D. 200

2 Which of the following does NOT belong to non-power benefits of hydropower?
   A. Water supply.   B. Flood control.
   C. Flexible electricity.   D. Irrigation.

3 Hydropower plants capture the energy of _____ to generate electricity.
   A. rising water   B. falling water
   C. turbine   D. reservoir

4 In a hydroelectric plant, generator is connected to the _____ by shafts and possibly gears.
   A. dam   B. reservoir
   C. channel   D. turbine

5 All power plants may use the same system of _____ and stations in an area to bring power to you.
   A. transmission lines   B. transformers
   C. generators   D. reservoirs

V. **Translate the following sentences into Chinese.**
1. By upgrading and increasing the efficiencies and capacities of existing facilities, hydropower can continue to support the nation's growing energy needs.
2. Hydropower plants range in size from small systems for a home or village to large projects producing electricity for utilities.
3. When the demand for electricity is low, a pumped storage facility stores energy by pumping water from a lower reservoir to an upper reservoir.
4. Water released from the reservoir flows through a turbine, spinning it, which in turn activates a generator to produce electricity.
5. By use of this "power grid," electricity can be interchanged among several utility systems to meet varying demands.

VI. **Translate the following sentences into English.**
1. 水电可以通过水循环再生，是一种可靠的零排放资源。(emission-free)
2. 尽管水力发电具有许多优势，但该行业还面临着独特的环境挑战。(hydroelectric production)
3. 一个微型水电站的容量可达100千瓦。(a capacity of)
4. 大坝通常位于偏远地区，电力必须通过一定距离传输给用户。(transmit to)
5. 在当地的变电站，变压器降低电压，这样电力就可以在整个地区进行分配和传导。(be divided up)

VII. **Translate the passage into Chinese.**
The most common type of hydroelectric power plant is an impoundment facility. An impoundment facility, typically a large hydropower system, uses a dam to store river water in a reservoir. Water released from the reservoir flows through a turbine, spinning it, which in turn activates a generator to produce electricity. The water may be released either to meet changing electricity needs or to maintain a constant reservoir level.

Text B

Hydroelectric Energy and Its Pros and Cons

1   A hydroelectric power station converts the kinetic, or movement, energy in flowing or falling water into electrical energy that can be used in homes and businesses. Hydroelectric energy can be generated on a small scale with a "run-of-river" installation, which uses naturally flowing river water to turn one or more turbines, or on a large scale with a hydroelectric dam.

2   A hydroelectric dam straddles a river, blocking the water's progress downstream. Water collects on the upstream side of the dam, forming an artificial lake known as a reservoir. Damming the river converts the water's kinetic energy into potential energy: the reservoir becomes a sort of battery, storing energy that can be released a little at a time. As well as being a source of energy, some reservoirs are used as boating lakes or drinking water supplies.

3   The reservoir's potential energy is converted back into kinetic energy by opening underwater gates, or intakes, in the dam. When an intake opens, the immense weight of the reservoir forces water through a channel called the penstock towards a turbine. The water rushes past the turbine, hitting its blades and causing it to spin, converting some of the water's kinetic energy into mechanical energy. The water then finally flows out of the dam and continues its journey downstream. A shaft connects the turbine to a generator, so when the turbine spins, so does the generator. The generator uses an electromagnetic field to convert this mechanical energy into electrical energy.

4   As long as there is plenty of water in the reservoir, a hydroelectric dam can respond quickly to changes in demand for electricity. Opening and closing the intakes directly controls the amount of water flowing through the penstock, which determines the amount of electricity the dam is generating.

5   The turbine and generator are located in the dam's power house, which also houses a transformer. The transformer converts the electrical energy from the generator to a high voltage. The national grid uses high voltages to transmit electricity efficiently through the

power lines to the homes and businesses that need it. Here, other transformers reduce the voltage back down to a usable level.

Water Generates Electricity

6 So, how exactly is electricity generated from water? This usually occurs through a hydroelectric dam that is built along a flowing waterway. Think of the water at Niagara. The water flowing over the falls is traveling faster than the water approaching the falls. As water travels downhill, it picks up speed and power.

7 To take advantage of this, dams are constructed along the waterway where there is a large elevation drop. This is exactly why you wouldn't expect to find dams in flat places like Florida, but would expect to find them in the hilly, mountainous regions of the Southwest U.S.

8 Like going down a slide at the playground, water is pulled downhill by gravity and picks up speed as it goes (you're moving faster at the bottom of the slide than at the top, right?). Inside the dam are turbines that get spun by the moving water—similar to how a pinwheel gets spun by the wind blowing by.

9 The spinning turbine shafts are connected to a generator, and the spinning of the turbines themselves creates electricity inside the generator. The generator is connected to power lines, which transmit the electricity to homes and buildings just like they do with coal-fired and natural gas power plants.

10 Just like wind blowing past the pinwheel is not affected by the pinwheel itself, water flowing by the turbines is not affected as it passes through the dam. It flows on downstream as if nothing has happened. You can see why this is such a valuable and widely used energy resource all around the world! However, with every source of energy, there are pros and cons. Let's look at the benefits and drawbacks that come with harnessing the amazing power of moving water.

Hydroelectric Energy: Pros

1. Renewable

11 Hydroelectric energy is renewable. This means that we cannot use up. However, there's only a limited number of suitable reservoirs where hydroelectric power plants can be built and even less places where such projects are profitable.

2. Clean

12   Unlike fossil fuels, biomass and nuclear power, hydroelectric power is a clean and green alternative source of energy. Since hydroelectric dams do not use fuel, they do not release any greenhouse gases or toxins into the environment. As a result, hydroelectric power features prominently in the clean energy plans of many countries.

3. Safe

13   Compared to fossil fuels and nuclear energy, hydroelectricity is much safer. There is no fuel involved (other than water that is).

4. Other uses

14   Reservoirs created by hydroelectric projects often become tourist attractions in their own right. The lake that forms behind the dam can be used for recreational purposes such as water sports and leisure activities such as fishing and boating, and the lake's water can also be used for irrigation and aquaculture purposes. Besides, large hydroelectric power plants can control floods as they can store vast quantities of water.

Hydroelectric Energy: Cons

1. Ecosystem damage and loss of wetlands

15   Big reservoirs associated with traditional hydroelectric power plants cause submersion of extensive areas upstream of the dams, sometimes destroying lowland and riverine valley forests, marshland and grasslands.

16   Hydroelectric power plants can also spell doom to surrounding aquatic ecosystems both upstream and downstream of the plant site. Since turbine gates are often opened intermittently, interruptions of natural water flow can have a great impact on the river ecosystem and the environment. The fish in the river can be affected by the draining of the water from the dam as well as the fish that is in the dam. Animals such as birds, cranes and other aquatic birds, and some plant species thrive in marshy habitats. However, because of the hydroelectric power plant construction, these habitats will be destroyed.

2. Expensive

17   Hydroelectric power plants and dams can be incredibly expensive to construct, regardless of the type of building, due to logistical challenges. Moreover, the projects take long

periods to finish and will have to operate for a long time too to recover the money spent on construction.

3. Droughts

18 One of the main downsides of setting up hydroelectric power plants is the occurrence of local droughts. The overall cost of energy is calculated depending on the availability of water and a drought could potentially affect this, causing people not to acquire the power they need.

4. Relocation due to risk of floods

19 Local populations living downstream can become vulnerable to flooding due to the possible strong water currents that might be released from the dams. As a result, people are forced to relocate to facilitate the construction of the dams needed to generate hydroelectricity. The World Commission on Dams estimated in 2000 that dams had physically displaced 40-80 million people worldwide.

(1,136 words)

New Words

1. straddle /ˈstrædl/ v. 横跨
2. intake /ˈɪnteɪk/ n. 入口
3. pinwheel /ˈpɪnwiːl/ n. 风车
4. biomass /ˈbaɪəʊmæs/ n. 生物燃料
5. toxin /ˈtɒksɪn/ n. 毒素
6. aquaculture /ˈækwəkʌltʃə/ n. 水产业
7. riverine /ˈrɪvəraɪn/ adj. 河的
8. marshland /ˈmɑːʃlænd/ n. 沼泽地
9. intermittently /ˌɪntəˈmɪtəntli/ adv. 间歇地
10. logistical /ləˈdʒɪstɪkəl/ adj. 后勤方面的
11. downside /ˈdaʊnsaɪd/ n. 消极面
12. relocate /ˌriːləʊˈkeɪt/ v. 重新安置
13. displace /dɪsˈpleɪs/ v. 移置

Unit 9　Hydroelectric Power

Expressions

1	pros and cons	利弊
2	electromagnetic field	电磁场
3	power house	发电站
4	high voltage	高电压
5	national grid	国家电网
6	power line	输电线

EXERCISES

I. Read each of the following paraphrases and then write the word it represents on the line provided.

1　a disadvantage or inconvenience _____

2　to move to or establish in a new place _____

3　toward or closer to the source of a stream _____

4　a habitat that is dominated by marshes, swamps, bogs, and the like_____

5　consisting of, relating to, or being in water _____

II. Find the Chinese equivalents for the following expressions.

1　electromagnetic field

2　high voltage

3　national grid

4　power line

5　elevation drop

III. Discuss the following questions.

1　What can reservoirs be used for besides being as a source of energy?

2　How can a reservoir's potential energy be converted back into kinetic energy?

3　How can the amount of water flowing through the penstock be controlled?

4　What's the function of a transformer in a power station?

5　Where are dams usually constructed? Why?

IV. Decide whether the following statements are true or false according to the text.

____ 1 Hydroelectricity is a kind of renewable energy that cannot be used up.

____ 2 There are not as many suitable reservoirs as expected where hydroelectric power plants can be built.

____ 3 The amount of electricity a dam generates is non-adjustable because of the steady water flow.

____ 4 Big reservoirs associated with traditional hydroelectric power plants are benign to ecosystem.

____ 5 The habitats of some aquatic birds may be destroyed by the construction of hydroelectric power plants.

V. Translate the following sentences into English.

1 国家电网利用高压输电线向居民和企业高效输电。(transmit)

2 当水往下流动时，速度和动力都加大了。(pick up)

3 大坝通常沿着水位下降较大的水道建造。(elevation drop)

4 水电站不使用燃料，因此不会把温室气体和有毒物质排放到周围环境。(release)

5 大坝的发电量可能受到当地旱情的影响。(local droughts)

VI. Translate the passage into Chinese.

The reservoir's potential energy is converted back into kinetic energy by opening underwater gates, or intakes, in the dam. When an intake opens, the immense weight of the reservoir forces water through a channel called the penstock towards a turbine. The water rushes past the turbine, hitting its blades and causing it to spin, converting some of the water's kinetic energy into mechanical energy. The water then finally flows out of the dam and continues its journey downstream. A shaft connects the turbine to a generator, so when the turbine spins, so does the generator. The generator uses an electromagnetic field to convert this mechanical energy into electrical energy.

Unit 10
Water Pollution

Lead-in

Water pollution is the release of substances into subsurface groundwater or into lakes, streams, rivers, estuaries, and oceans to the point where the substances interfere with beneficial use of the water or with the natural functioning of ecosystems. In addition to the release of substances, such as chemicals or microorganisms, water pollution may also include the release of energy, in the form of radioactivity or heat, into bodies of water.

Text A

Water Pollution—Everything You Need to Know

1 The widespread problem of water pollution is jeopardizing our health. Unsafe water kills more people each year than war and all other forms of violence combined. Meanwhile, our drinkable water sources are finite: Less than 1 percent of the Earth's freshwater is actually accessible to us. Without action, the challenges will only increase by 2050, when global demand for freshwater is expected to be one-third greater than it is now.

What Is Water Pollution?

2 Water pollution occurs when harmful substances—often chemicals or microorganisms—contaminate a stream, river, lake, ocean, aquifer, or other body of water, degrading water quality and rendering it toxic to humans or the environment.

What Are the Causes of Water Pollution?

3 Water is uniquely vulnerable to pollution. Known as a "universal solvent", water is able to dissolve more substances than any other liquid on Earth. It's also why water is so easily polluted. Toxic substances from farms, towns, and factories readily dissolve into and mix with it, causing water pollution.

Groundwater

4 When rain falls and seeps deep into the earth, filling the cracks, crevices, and porous spaces of an aquifer (basically an underground storehouse of water), it becomes groundwater—one of our least visible but most important natural resources. Nearly 40 percent of Americans rely on groundwater, pumped to the Earth's surface, for drinking water. For some folks in rural areas, it's their only freshwater source. Groundwater gets polluted when contaminants—from pesticides and fertilizers to waste leached from landfills and septic systems—make their way into an aquifer, rendering it unsafe for human use. Ridding groundwater of contaminants can be difficult to impossible, as well as costly. Once polluted, an aquifer may be unusable for decades, or even thousands of

years. Groundwater can also spread contamination far from the original polluting source as it seeps into streams, lakes, and oceans.

Surface water

5   Covering about 70 percent of the Earth, surface water is what fills our oceans, lakes, rivers, and all those other blue bits on the world map. Surface water from freshwater sources (that is, from sources other than the ocean) accounts for more than 60 percent of the water delivered to American homes. But a significant pool of that water is in peril. According to the most recent surveys on national water quality from the U.S. Environmental Protection Agency, nearly half of our rivers and streams and more than one-third of our lakes are polluted and unfit for swimming, fishing, and drinking. Nutrient pollution, which includes nitrates and phosphates, is the leading type of contamination in these freshwater sources. While plants and animals need these nutrients to grow, they have become a major pollutant due to farm waste and fertilizer runoff. Municipal and industrial waste discharges contribute their fair share of toxins as well. There's also all the random junk that industry and individuals dump directly into waterways.

Ocean water

6   Eighty percent of ocean pollution (also called marine pollution) originates on land—whether along the coast or far inland. Contaminants such as chemicals, nutrients, and heavy metals are carried from farms, factories, and cities by streams and rivers into our bays and estuaries; from there they travel out to sea. Meanwhile, marine debris—particularly plastic—is blown in by the wind or washed in via storm drains and sewers. Our seas are also sometimes spoiled by oil spills and leaks—big and small—and are consistently soaking up carbon pollution from the air. The ocean absorbs as much as a quarter of man-made carbon emissions.

Point source

7   When contamination originates from a single source, it's called point source pollution. Examples include wastewater (also called effluent) discharged legally or illegally by a manufacturer, oil refinery, or wastewater treatment facility, as well as contamination from leaking septic systems, chemical and oil spills, and illegal dumping. The EPA regulates point source pollution by establishing limits on what can be discharged by a facility directly into a body of water. While point source pollution originates from a specific

place, it can affect miles of waterways and ocean.

Nonpoint source

8   Nonpoint source pollution is contamination derived from diffuse sources. These may include agricultural or stormwater runoff or debris blown into waterways from land. Nonpoint source pollution is the leading cause of water pollution in U.S. waters, but it's difficult to regulate, since there's no single, identifiable culprit.

Transboundary

9   It goes without saying that water pollution can't be contained by a line on a map. Transboundary pollution is the result of contaminated water from one country spilling into the waters of another. Contamination can result from a disaster—like an oil spill—or the slow, downriver creep of industrial, agricultural, or municipal discharge.

The Most Common Types of Water Contamination

Agricultural

10 Not only is the agricultural sector the biggest consumer of global freshwater resources, with farming and livestock production using about 70 percent of the Earth's surface water supplies, but it's also a serious water polluter. Around the world, agriculture is the leading cause of water degradation. In the United States, agricultural pollution is the top source of contamination in rivers and streams, the second-biggest source in wetlands, and the third main source in lakes. It's also a major contributor of contamination to estuaries and groundwater. Every time it rains, fertilizers, pesticides, and animal waste from farms and livestock operations wash nutrients and pathogens—such bacteria and viruses—into our waterways. Nutrient pollution, caused by excess nitrogen and phosphorus in water or air, is the number-one threat to water quality worldwide and can cause algal blooms, a toxic soup of blue-green algae that can be harmful to people and wildlife.

Sewage and wastewater

11 Used water is wastewater. It comes from our sinks, showers, and toilets (think sewage) and from commercial, industrial, and agricultural activities (think metals, solvents, and toxic sludge). The term also includes stormwater runoff, which occurs when rainfall carries road salts, oil, grease, chemicals, and debris from impermeable surfaces into our waterways.

12   More than 80 percent of the world's wastewater flows back into the environment without being treated or reused, according to the United Nations; in some least-developed countries, the figure tops 95 percent. In the United States, wastewater treatment facilities process about 34 billion gallons of wastewater per day. These facilities reduce the amount of pollutants such as pathogens, phosphorus, and nitrogen in sewage, as well as heavy metals and toxic chemicals in industrial waste, before discharging the treated waters back into waterways. That's when all goes well. But according to EPA estimates, our nation's aging and easily overwhelmed sewage treatment systems also release more than 850 billion gallons of untreated wastewater each year.

Oil pollution

13   Big spills may dominate headlines, but consumers account for the vast majority of oil pollution in our seas, including oil and gasoline that drips from millions of cars and trucks every day. Moreover, nearly half of the estimated 1 million tons of oil that makes its way into marine environments each year comes not from tanker spills but from land-based sources such as factories, farms, and cities. At sea, tanker spills account for about 10 percent of the oil in waters around the world, while regular operations of the shipping industry—through both legal and illegal discharges—contribute about one-third. Oil is also naturally released from under the ocean floor through fractures known as seeps.

Radioactive substances

14   Radioactive waste is any pollution that emits radiation beyond what is naturally released by the environment. It's generated by uranium mining, nuclear power plants, and the production and testing of military weapons, as well as by universities and hospitals that use radioactive materials for research and medicine. Radioactive waste can persist in the environment for thousands of years, making disposal a major challenge. Consider the decommissioned Hanford nuclear weapons production site in Washington, where the cleanup of 56 million gallons of radioactive waste is expected to cost more than $100 billion and last through 2060. Accidentally released or improperly disposed of contaminants threaten groundwater, surface water, and marine resources.

(1,325 words)

New Words

1. estuary /ˈestʃuəri/ n.　　　　　河口
2. microorganism /ˌmaɪkrəʊˈɔːɡənɪzəm/ n.　　微生物
3. jeopardize /ˈdʒepədaɪz/ v.　　　危及
4. septic /ˈseptɪk/ adj.　　　　　腐败的
5. peril /ˈperɪl/ n.　　　　　　　危险
6. nitrate /ˈnaɪtreɪt/ n.　　　　　硝酸盐
7. phosphate /ˈfɒsfeɪt/ n.　　　　磷酸盐
8. dump /dʌmp/ v.　　　　　　　倾倒
9. marine /məˈriːn/ adj.　　　　　海的
10. diffuse /dɪˈfjuːs/ adj.　　　　弥漫的
11. stormwater /ˈstɔːmˌwɔːtə/ n.　　暴雨水
12. culprit /ˈkʌlprɪt/ n.　　　　　起因
13. degradation /ˌdeɡrəˈdeɪʃən/ n.　退化
14. sludge /slʌdʒ/ n.　　　　　　淤泥
15. impermeable /ɪmˈpɜːmiəbəl/ adj.　不能渗透的
16. decommission /ˌdiːkəˈmɪʃən/ v.　停止使用

Expressions

1. septic system　　　　　　　化粪池系统
2. nutrient pollution　　　　　营养物污染
3. heavy metal　　　　　　　重金属
4. storm drain　　　　　　　雨水管道
5. carbon emission　　　　　碳排放量
6. point source pollution　　　点源污染
7. oil refinery　　　　　　　炼油厂
8. blue-green algae　　　　　蓝绿藻
9. toxic chemical　　　　　　有毒化学品
10. marine environment　　　　海洋环境
11. radioactive substance　　　放射性物质

> **Note**
>
> **U.S. Environmental Protection Agency** 美国国家环境保护局，是美国联邦政府的一个独立行政机构，主要负责维护自然环境和保护人类健康不受环境危害影响。

Exercises

I. Read each of the following paraphrases and then write the word it represents on the line provided.

1. to put in danger; imperil _____
2. a substance, usually a liquid, capable of dissolving another substance _____
3. something that endangers or involves risk _____
4. of or relating to the sea _____
5. to cause to spread out freely _____
6. the reason for a particular problem or difficulty _____
7. a decline to a lower condition, quality, or level _____
8. to release or throw down in a large mass _____
9. not allowing the passage of a fluid through _____
10. to take (a ship or a military base, for example) out of active service or use; render inactive _____

II. Find the Chinese equivalents for the following expressions.

1. nutrient pollution
2. heavy metal
3. storm drain
4. carbon emission
5. point source pollution
6. oil refinery
7. blue-green algae
8. toxic chemical
9. marine environment
10. radioactive substance

III. Discuss the following questions.

1. What is water pollution?

2 Why is water so easily polluted?

3 What are the differences between the point source pollution and nonpoint source pollution?

4 What is the biggest threat to water quality in the globe?

5 What are the easily polluted water sources and the most common types of water contamination?

IV. Choose the best answer to each of the following questions or statements.

1 What is one of the least visible but most important natural resources mentioned in the passage?
   A. Storm water.   B. Ocean water.
   C. Surface water.   D. Groundwater.

2 If the rivers, streams and lakes are polluted, the water may be fit for _____.
   A. swimming   B. washing
   C. drinking   D. fishing

3 Which of the following pollution sources is difficult to regulate?
   A. Ocean water.   B. Transboundary.
   C. Point source.   D. Nonpoint source.

4 Which of the following CANNOT be reduced by wastewater treatment facilities?
   A. Uranium.   B. Pathogen.
   C. Phosphoru.   D. Nitrogen.

5 If you were asked to write a continuation of this passage, which part would you decide to write?
   A. Harm of water pollution.
   B. Human factors causing water pollution.
   C. How to control water pollution.
   D. How to avoid water pollution.

V. Translate the following sentences into Chinese.

1 Toxic substances from farms, towns, and factories readily dissolve into and mix with water, causing water pollution.

2 Nutrient pollution, which includes nitrates and phosphates, is the leading type of contamination in these freshwater sources.

3 Nutrient pollution, caused by excess nitrogen and phosphorus in water or air, is the number-one threat to water quality worldwide and can cause algal blooms, a toxic

soup of blue-green algae that can be harmful to people and wildlife.

4 These facilities reduce the amount of pollutants such as pathogens, phosphorus, and nitrogen in sewage, as well as heavy metals and toxic chemicals in industrial waste, before discharging the treated waters back into waterways.

5 Radioactive waste is any pollution that emits radiation beyond what is naturally released by the environment.

VI. Translate the following sentences into English.
1. 水比地球上其他任何液体都更能溶解物质，因此被称为"万能溶剂"。(dissolve)
2. 海洋废弃物，特别是塑料，是被风吹进海里的，或者是通过雨水道和下水道冲入海洋中的。(storm drain)
3. 非点源污染是源于扩散源的污染。(derive from)
4. 跨界污染是由被污染的水从一个国家流到另一个国家水域所造成的。(transboundary)
5. 石油也会通过海底的裂缝自然释放出来。(seep)

VII. Translate the passage into Chinese.

The widespread problem of water pollution is jeopardizing our health. Unsafe water kills more people each year than war and all other forms of violence combined. Meanwhile, our drinkable water sources are finite: Less than 1 percent of the Earth's freshwater is actually accessible to us. Without action, the challenges will only increase by 2050, when global demand for freshwater is expected to be one-third greater than it is now.

Text B

What Are the Effects of Water Pollution?

On human health

1 To put it bluntly: Water pollution kills. In fact, it caused 1.8 million deaths in 2015, according to a study published in *The Lancet*. Contaminated water can also make you ill. Every year, unsafe water sickens about 1 billion people. And low-income communities are disproportionately at

risk because their homes are often closest to the most polluting industries.

2. Waterborne pathogens, in the form of disease-causing bacteria and viruses from human and animal waste, are a major cause of illness from contaminated drinking water. Diseases spread by unsafe water include cholera, giardia, and typhoid. Even in wealthy nations, accidental or illegal releases from sewage treatment facilities, as well as runoff from farms and urban areas, contribute harmful pathogens to waterways. Thousands of people across the United States are sickened every year by Legionnaires' disease (a severe form of pneumonia contracted from water sources like cooling towers and piped water), with cases cropping up from California's Disneyland to Manhattan's Upper East Side.

3. Meanwhile, the plight of residents in Flint, Michigan—where cost-cutting measures and aging water infrastructure created the recent lead contamination crisis—offers a stark look at how dangerous chemical and other industrial pollutants in our water can be. The problem goes far beyond Flint and involves much more than lead, as a wide range of chemical pollutants—from heavy metals such as arsenic and mercury to pesticides and nitrate fertilizers—are getting into our water supplies. Once they're ingested, these toxins can cause a host of health issues, from cancer to hormone disruption to altered brain function. Children and pregnant women are particularly at risk.

4. Even swimming can pose a risk. Every year, 3.5 million Americans contract health issues such as skin rashes, pinkeye, respiratory infections, and hepatitis from sewage-laden coastal waters, according to EPA estimates.

On the environment

5. In order to thrive, healthy ecosystems rely on a complex web of animals, plants, bacteria, and fungi—all of which interact, directly or indirectly, with each other. Harm to any of these organisms can create a chain effect, imperiling entire aquatic environments.

6. When water pollution causes an algal bloom in a lake or marine environment, the proliferation of newly introduced nutrients stimulates plant and algae growth, which in turn reduces oxygen levels in the water. This dearth of oxygen, known as eutrophication, suffocates plants and animals and can create "dead zones," where waters are essentially devoid of life. In certain cases, these harmful algal blooms can also produce neurotoxins that affect wildlife, from whales to sea turtles.

7   Chemicals and heavy metals from industrial and municipal wastewater contaminate waterways as well. These contaminants are toxic to aquatic life—most often reducing an organism's life span and ability to reproduce—and make their way up the food chain as predator eats prey. That's how tuna and other big fish accumulate high quantities of toxins, such as mercury.

8   Marine ecosystems are also threatened by marine debris, which can strangle, suffocate, and starve animals. Much of this solid debris, such as plastic bags and soda cans, gets swept into sewers and storm drains and eventually out to the sea, turning our oceans into trash soup and sometimes consolidating to form floating garbage patches. Discarded fishing gear and other types of debris are responsible for harming more than 200 different species of marine life.

9   Meanwhile, ocean acidification is making it tougher for shellfish and coral to survive. Though they absorb about a quarter of the carbon pollution created each year by burning fossil fuels, oceans are becoming more acidic. This process makes it harder for shellfish and other species to build shells and may impact the nervous systems of sharks, clownfish, and other marine life.

What can you do to prevent water pollution?

With your actions

10 It's easy to tsk-tsk the oil company with a leaking tanker, but we're all accountable to some degree for today's water pollution problem. Fortunately, there are some simple ways you can prevent water contamination or at least limit your contribution to it:

- Reduce your plastic consumption and reuse or recycle plastic when you can.
- Properly dispose of chemical cleaners, oils, and non-biodegradable items to keep them from ending up down the drain.
- Maintain your car so it doesn't leak oil, antifreeze, or coolant.
- If you have a yard, consider landscaping that reduces runoff and avoid applying pesticides and herbicides.
- If you have a pup, be sure to pick up its poop.

With your voice

11 One of the most effective ways to stand up for our waters is to speak out in support of the Clean Water Rule, which clarifies the Clean Water Act's scope and protects the drinking water of one in three Americans.

12 Tell the federal government, the U.S. Army Corps of Engineers, and your local elected officials that you support the Clean Water Rule. Also, learn how you and those around you can get involved in the policy-making process. Our public waterways serve every American. We should all have a say in how they're protected.

(833 words)

New Words

1	bluntly	/'blʌntli/ ad.	直率地
2	waterborne	/'wɔːtəbɔːn/ adj.	水传播的
3	cholera	/'kɒlərə/ n.	霍乱
4	giardia	/dʒiː'ɑːdiə/ n.	鞭毛虫
5	typhoid	/'taɪfɔɪd/ n.	伤寒
6	pneumonia	/njuː'məʊniə/ n.	肺炎
7	plight	/plaɪt/ n.	困境
8	arsenic	/'ɑːsənɪk/ n.	砷
9	mercury	/'mɜːkjʊri/ n.	水银
10	hormone	/'hɔːməʊn/ n.	激素
11	pinkeye	/'pɪŋkaɪ/ n.	传染性结膜炎
12	hepatitis	/ˌhepə'taɪtɪs/ n.	肝炎
13	thrive	/θraɪv/ v.	繁荣，兴旺
14	fungi	/'fʌŋgaɪ/ n. pl.	霉菌，菌类
15	proliferation	/prəˌlɪfə'reɪʃn/ n.	增殖，扩散
16	dearth	/dɜːθ/ n.	缺乏
17	eutrophication	/ˌjuːtrəfɪ'keɪʃn/ n.	富营养化
18	suffocate	/'sʌfəkeɪt/ v.	使缺氧

19	neurotoxin /ˌnjuərəuˈtɒksɪn/ n.	神经毒素
20	predator /ˈpredətə/ n.	捕食者
21	strangle /ˈstræŋɡəl/ v.	扼死
22	starve /stɑːv/ v.	挨饿
23	non-biodegradable /ˌnɒnˌbaɪəudɪˈɡreɪdəbl/ adj.	不能进行生物降解的
24	antifreeze /ˈæntɪfriːz/ n.	防冻剂
25	coolant /ˈkuːlənt/ n.	冷冻剂

Expressions

1	piped water	自来水
2	lead contamination	铅污染
3	nitrate fertilizer	硝酸盐肥料
4	hormone disruption	激素紊乱
5	respiratory infection	呼吸道感染
6	chain effect	连锁效应
7	marine ecosystem	海洋生态系
8	fishing gear	钓鱼用具
9	ocean acidification	海洋酸化
10	fossil fuel	矿物燃料

Notes

1 **The Lancet**《柳叶刀》，创刊于1823年，是由爱思唯尔（Elsevier）出版公司主办、《柳叶刀》编辑部编辑出版的医学学术刊物，是国际公认的综合性医学四大期刊之一。

2 **Legionnaires' disease** 军团菌病，是由军团菌属细菌引起的临床综合征，因1976年美国费城召开退伍军人大会时暴发而得名。病原菌主要来自土壤和污水，由空气传播，自呼吸道侵入。

3 **Manhattan's Upper East Side** 曼哈顿上东区，毗邻中央公园，汇集了世界著名的博物馆和顶尖名牌购物中心，如大都会艺术博物馆等。

EXERCISES

I. Read each of the following paraphrases and then write the word it represents on the line provided.

1. spread or carried by water _____
2. to kill or destroy by preventing access of air or oxygen _____
3. to grow or develop vigorously; flourish _____
4. the over-enrichment of water by nutrients, causing overgrowth and decay of plants, deoxygenation of water, and the death of its organisms _____
5. an organism that lives by preying on other organisms _____

II. Find the Chinese equivalents for the following expressions.

1. piped water
2. lead contamination
3. nitrate fertilizer
4. hormone disruption
5. respiratory infection
6. chain effect
7. marine ecosystem
8. fishing gear
9. ocean acidification
10. fossil fuel

III. Discuss the following questions.

1. What is the main reason for illness from polluted drinking water?
2. What chemical pollutants can cause water pollution?
3. What harm could swimming in the polluted water do to people?
4. How do tuna and other big fish accumulate a lot of toxins?
5. What can we do to prevent water pollution?

IV. Decide whether the following statements are true or false according to the text.

_____ 1 Diseases spread by unsafe water include cholera, giardia, and typhoid.
_____ 2 Children and pregnant women are more vulnerable to chemical contamination.
_____ 3 In order to thrive, healthy ecosystems rely on a complex web of animals, plants, bacteria, and virus.

_____ 4 Eutrophication suffocates plants and animals and can create "dead zones".

_____ 5 Heavy metal pollution will affect the life span and reproductive capacity of aquatic organisms.

_____ 6 Solid waste which is washed into sewers does not pollute the ocean.

_____ 7 Ocean acidification is making it tougher for shellfish and coral to survive.

_____ 8 Every one of us, to some extent, should shoulder the responsibility of fighting water pollution facing us nowadays.

_____ 9 Supporting the Clean Water Rule is the only way to protect water resources for Americans.

_____ 10 In the United States, only the federal government, the U.S. Army Corps of Engineers, and local elected officials have the right to speak up for the protection of clean water.

V. **Translate the following sentences into English.**

1 一旦摄入这些毒素，就会给宿主造成许多健康问题。(host)
2 任何一种生物受到伤害都会产生连锁反应，从而危及整个水生环境。(chain effect)
3 工业和城市污水中的化学物质和重金属也会污染水道。(contaminate)
4 海洋生态系统也受到海洋废弃物的威胁，使得动物缺氧窒息并遭受饥饿。(threaten)
5 尽管它们每年吸收约四分之一的化石燃料所产生的碳污染，但海洋的酸性依然与日俱增。(absorb)

VI. **Translate the passage into Chinese.**

When water pollution causes an algal bloom in a lake or marine environment, the proliferation of newly introduced nutrients stimulates plant and algae growth, which in turn reduces oxygen levels in the water. This dearth of oxygen, known as eutrophication, suffocates plants and animals and can create "dead zones", where waters are essentially devoid of life. In certain cases, these harmful algal blooms can also produce neurotoxins that affect wildlife, from whales to sea turtles.

Appendix
Glossary

A

abiotic /ˌeɪbaɪˈɒtɪk/ adj. 非生物的
abutment /əˈbʌtmənt/ n. 拱基
accumulate /əˈkjuːmjəleɪt/ v. 积累，积聚
acreage /ˈeɪkərɪdʒ/ n. 大块土地
activate /ˈæktɪveɪt/ v. 激活
address /əˈdres/ v. 处理
aerial /ˈeəriəl/ adj. 空中的
aftermath /ˈɑːftəmæθ/ n. 后果，余波
aggravate /ˈægrəveɪt/ v. 恶化
algae /ˈældʒiː/ n. 水藻；海藻
alkaline /ˈælkəlaɪn/ adj. 含碱的，碱性的
alleviate /əˈliːvieɪt/ v. 减轻，使……缓和
aluminum /æljəˈmɪnəm/ n. 铝
ancillary /ænˈsɪləri/ adj. 辅助的
antifreeze /ˈæntɪfriːz/ n. 防冻剂
aquaculture /ˈækwəkʌltʃə/ n. 水产业
aquatic /əˈkwætɪk/ adj. 水生的
aquifer /ˈækwɪfə/ n. 含水土层
aquifer /ˈækwɪfə/ n. 含水层
arch /ɑːtʃ/ n. 拱，拱状物
argon /ˈɑːgɒn/ n. 氩
arid /ˈærɪd/ adj. 干旱的
arsenic /ˈɑːsənɪk/ n. 砷
auxiliary /ɔːgˈzɪljəri/ adj. 辅助的，附加的
axial /ˈæksiəl/ adj. 轴的

B

barrage /ˈbærɑːʒ/ n. 拦河坝
bedrock /ˈbedrɒk/ n. 基岩（松软的沙、土层下的岩石）
benign /bɪˈnaɪn/ adj. 良性的
berm /bɜːm/ n. 护坡道
biomass /ˈbaɪəʊmæs/ n. 生物燃料
bluntly /ˈblʌntli/ adv. 直率地
bog /bɒg/ n. 沼泽
brace /breɪs/ v. 支撑

Appendix Glossary

bund /bʌnd/ n. 堤岸，河岸
buttress /ˈbʌtrəs/ n. 拱壁

C

capillary /kəˈpɪləri/ n. 毛细管
cavitation /ˌkævɪˈteɪʃən/ n. 气蚀
cholera /ˈkɒlərə/ n. 霍乱
chute /ʃuːt/ n. 斜槽，溜槽
circulation /ˌsɜːkjəˈleɪʃən/ n. 循环
cistern /ˈsɪstən/ n. 蓄水池
clay /kleɪ/ n. 黏土
cofferdam /ˈkɒfədæm/ n. 围堰
collapse /kəˈlæps/ v. 倒塌，崩溃，瓦解
compacted /kəmˈpæktɪd/ adj. 夯实的，紧实的
compensate /ˈkɒmpənseɪt/ v. 补偿
compound /kəmˈpaʊnd/ v. 使恶化
concomitant /kənˈkɒmɪtənt/ adj. 相伴的，附随的
concrete /ˈkɒŋkriːt/ n. 水泥，混凝土
condense /kənˈdens/ v. 冷凝
cone /kəʊn/ n. 圆锥体，锥形物
confine /kənˈfaɪn/ v. 限制，局限
confront /kənˈfrʌnt/ v. 面临
congregate /ˈkɒŋɡrɪɡeɪt/ v. 聚集
containment /kənˈteɪnmənt/ n. 控制
contaminant /kənˈtæmɪnənt/ n. 杂质，污染物
contaminate /kənˈtæmɪneɪt/ v. 污染
convert /kənˈvɜːt/ v. 转化
coolant /ˈkuːlənt/ n. 冷冻剂
counteraction /ˌkaʊntərˈækʃən/ n. 中和，抵消
creek /kriːk/ n. 小湾，小溪
crest /krest/ n. 顶部
crowdsource /ˈkraʊdsɔːs/ n. 众包（指把工作任务外包给大众的做法）
crustacean /krʌˈsteɪʃən/ n. & adj. 甲壳类动物
culprit /ˈkʌlprɪt/ n. 起因

D

dearth /dɜːθ/ n. 缺乏
debris /ˈdebriː/ n. 碎片，残骸
decommission /ˌdiːkəˈmɪʃən/ v. 停止使用
deforestation /diːˌfɒrɪˈsteɪʃən/ n. 毁林，滥伐森林
degradation /ˌdeɡrəˈdeɪʃən/ n. 退化
deplete /dɪˈpliːt/ v. 耗尽
desalinate /diːˈsælɪneɪt/ v. 淡化（海水），（从海水中）除去盐分
devastating /ˈdevəsteɪtɪŋ/ adj. 毁灭性的
dew /djuː/ n. 露水

diffuse /dɪˈfjuːs/ adj.	弥漫的
dike /daɪk/ n.	堤坝
discrete /dɪˈskriːt/ adj.	不连续的，分离的
disperse /dɪˈspɜːs/ v.	分散，扩散
displace /dɪsˈpleɪs/ v.	移置
disposition /ˌdɪspəˈzɪʃn/ n.	排列
ditch /dɪtʃ/ n.	沟渠，壕沟
diversion /daɪˈvɜːʃn/ n.	引水
divert /daɪˈvɜːt/ v.	转移
downplay /ˌdaʊnˈpleɪ/ v.	低估
downside /ˈdaʊnsaɪd/ n.	消极面
drain /dreɪn/ v.	排出，排干
drainage /ˈdreɪnɪdʒ/ n.	排水
drastic /ˈdræstɪk/ adj.	激烈的
drawback /ˈdrɔːbæk/ n.	缺点
drill /drɪl/ v.	钻孔
drone /drəʊn/ n.	无人驾驶飞机
dump /dʌmp/ v.	倾倒
dyke /daɪk/ n.	堤坝

E

earthfill /ˈɜːθfɪl/ adj.	填土的
elevation /ˌelɪˈveɪʃn/ n.	海拔，高度
ellipse /ɪˈlɪps/ n.	椭圆
embankment /ɪmˈbæŋkmənt/ n.	堤防
embargo /ɪmˈbɑːgəʊ/ n.	禁运
emplace /ɪmˈpleɪs/ v.	安置就位
epilimnion /ˌepɪˈlɪmnɪən/ n.	湖面温水层
erosion /ɪˈrəʊʒən/ n.	腐蚀，侵蚀
estuary /ˈestʃuəri/ n.	河口
eutrophication /juːtrəfɪˈkeɪʃn/ n.	富营养化
evaporate /ɪˈvæpəreɪt/ v.	（使）蒸发，挥发
evaporation /ɪˌvæpəˈreɪʃn/ n.	蒸发
excavate /ˈekskəveɪt/ v.	挖掘，挖开
extraction /ɪkˈstrækʃn/ n.	抽取

F

fertilizer /ˈfɜːtɪlaɪzə/ n.	肥料
finitude /ˈfɪnɪtjuːd/ n.	有限性
fishery /ˈfɪʃəri/ n.	渔业
flame /fleɪm/ n.	火焰
floodgate /ˈflʌdgeɪt/ n.	泄水闸门，防洪闸门
floodplain /ˈflʌdpleɪn/ n.	漫滩
flora /ˈflɔːrə/ n.	植物，植物群落
fluctuation /ˌflʌktʃuˈeɪʃn/ n.	波动

Appendix Glossary

forage /ˈfɒrɪdʒ/ n.   草料，饲料
friction /ˈfrɪkʃən/ n.   摩擦力
fungi /ˈfʌŋgaɪ/ n. pl.   霉菌，菌类
furrow /ˈfʌrəʊ/ n.   畦沟

G

geology /dʒiˈɒlədʒi/ n.   （某地区的）地质情况
geothermal /ˌdʒiːəʊˈθɜːməl/ adj.   地热的
giardia /dʒiːˈɑːdiə/ n.   鞭毛虫
glacier /ˈglæsiə/ n.   冰川
goggle /ˈgɒgəl/ n.   护目镜
gravity /ˈgrævəti/ n.   重力
guesswork /ˈgeswɜːk/ n.   臆测，猜测

H

habitat /ˈhæbətæt/ n.   栖息地
halite /ˈhælaɪt/ n.   岩盐
harbor /ˈhɑːbə/ v.   包含
harness /ˈhɑːnəs/ v.   利用
hazard /ˈhæzəd/ n.   危险，危害
hepatitis /ˌhepəˈtaɪtəs/ n.   肝炎
heterotroph /ˈhetərətrəʊf/ n.   异养生物
hibernate /ˈhaɪbəneɪt/ v.   冬眠
hormone /ˈhɔːməʊn/ n.   激素
humidity /hjuːˈmɪdəti/ n.   湿度
hurricane /ˈhʌrəkən/ n.   飓风
hypolimnion /ˌhaɪpəʊˈlɪmnɪən/ n.   下层滞水带，湖底静水层
hyporheic /ˌhɪpəˈreɪk/ adj.   潜流的

I

imminently /ˈɪmɪnəntli/ adv.   迫切地，紧急地
impair /ɪmˈpeə/ v.   损害，削弱
impermeable /ɪmˈpɜːmiəbəl/ adj.   不能渗透的
impound /ɪmˈpaʊnd/ v.   蓄水
impoundment /ɪmˈpaʊndmənt/ n.   蓄水
incumbent /ɪnˈkʌmbənt/ adj.   有责任的
indelible /ɪnˈdeləbəl/ adj.   不能擦除的，不能磨灭的
infiltrate /ˈɪnfɪltreɪt/ v.   渗透
infrastructure /ˈɪnfrəˌstrʌktʃə/ n.   基础设施
install /ɪnˈstɔːl/ v.   安装
intake /ˈɪnteɪk/ n.   入口
integrity /ɪnˈtegrəti/ n.   完整
interfere /ˌɪntəˈfɪə/ v.   干扰，妨碍
intermittently /ˌɪntəˈmɪtəntli/ adv.   间歇地
intrusion /ɪnˈtruːʒən/ n.   入侵，闯入

inundate /ˈɪnəndeɪt/ v. 淹没
irrigation /ˌɪrɪˈgeɪʃən/ n. 灌溉

J

jeopardize /ˈdʒepədaɪz/ v. 危及
jurisdiction /ˌdʒʊərɪsˈdɪkʃən/ n. 行政辖区

K

kinetic /kɪˈnetɪk/ adj. 运动的

L

lagoon /ləˈguːn/ n. 小而浅的湖
larval /ˈlɑːvl/ n. 幼虫的，幼虫状态的
lateral /ˈlætərəl/ n. 侧面
layperson /ˈleɪˌpɜːsən/ n. 外行，非专业人员
levee /ˈlevi/ n. 防洪堤
leverage /ˈliːvərɪdʒ/ v. 利用
limnetic /lɪmˈnetɪk/ adj. 湖泊的
littoral /ˈlɪtərəl/ n. & adj. 海岸的，沿海的
livestock /ˈlaɪvstɒk/ n. pl. 家畜
logistical /ləˈdʒɪstɪkəl/ adj. 后勤方面的

M

magnitude /ˈmægnɪtjuːd/ n. 巨大
maintenance /ˈmeɪntənəns/ n. 维护
marine /məˈriːn/ adj. 海的
marsh /mɑːʃ/ n. 沼泽
marshal /ˈmɑːʃəl/ v. 整顿，配置
marshland /ˈmɑːʃlænd/ n. 沼泽地
masonry /ˈmeɪsənri/ n. 砖石结构
mayfly /ˈmeɪflaɪ/ n. 蜉蝣
meander /miˈændə/ n. 缓慢而弯曲地流动
mediate /ˈmiːdieɪt/ v. 调和
megawatt /ˈmegəwɒt/ n. 【电】兆瓦（特）
mentality /menˈtæləti/ n. 心态，思想方法
mercury /ˈmɜːkjʊri/ n. 水银
metalimnion /metəˈlɪmnɪən/ n. 变温层，斜温层
meteorological /ˌmiːtiərəˈlɒdʒɪkəl/ adj. 气象的
metropolitan /ˌmetrəˈpɒlɪtən/ adj. 大都市的
microorganism /ˌmaɪkrəʊˈɔːgənɪzəm/ n. 微生物
mine /maɪn/ v. 开采
miscellaneous /ˌmɪsəˈleɪniəs/ adj. 各种的
mitigate /ˈmɪtɪgeɪt/ v. 减轻，缓和
mitigation /ˌmɪtɪˈgeɪʃən/ n. 缓和，减轻
momentum /məʊˈmentəm/ n. 动量

monsoon /mɒnˈsuːn/ n. 季风；雨季
mudslide /ˈmʌdslaɪd/ n. 泥石流
multipurpose /ˌmʌltiˈpɜːpəs/ adj. 多种用途的，多目标的
municipal /mjuːˈnɪsɪpəl/ adj. 市政的
municipality /mjuːˌnɪsɪˈpæləti/ n. 直辖市
murky /ˈmɜːki/ adj. 混浊的

N

neurotoxin /ˌnjuərəʊˈtɒksɪn/ n. 神经毒素
nitrate /ˈnaɪtreɪt/ n. 硝酸盐
nitrogen /ˈnaɪtrədʒən/ n. 氮
non-biodegradable /ˌnɒnˌbaɪəʊdɪˈgreɪdəbl/ adj. 不能进行生物降解的
nourishment /ˈnʌrɪʃmənt/ n. 养料，营养
nozzle /ˈnɒzəl/ n. 喷嘴

O

ore /ɔː/ n. 矿石，矿物
outlet /ˈaʊtlet/ n. 排放孔
overload /ˌəʊvəˈləʊd/ v. 使超负荷
overtop /ˌəʊvəˈtɒp/ v. 漫溢

P

parallel /ˈpærəlel/ adj. 平行的
particulate /pɑːˈtɪkjəleɪt/ n. 微粒状物质
pathogen /ˈpæθədʒən/ n. 病原体
penstock /ˈpenstɒk/ n. 压力水管
percolate /ˈpɜːkəleɪt/ v. 渗入
perforate /ˈpɜːfəreɪt/ v. 穿孔
peril /ˈperəl/ n. 危险
permeable /ˈpɜːmiəbəl/ adj. 有渗透性的
perpendicular /ˌpɜːpənˈdɪkjʊlə/ adj. 垂直的
pesticide /ˈpestɪsaɪd/ n. 杀虫剂，农药
phosphate /ˈfɒsfeɪt/ n. 磷酸盐
phosphorus /ˈfɒsfərəs/ n. 磷
photosynthesis /ˌfəʊtəʊˈsɪnθəsɪs/ n. 光合作用
physiographic /ˈfɪzɪəˈgræfɪk/ adj. 地形学的
phytoplankton /ˈfaɪtəʊˌplæŋktən/ n. 浮游植物
pier /pɪə/ n. 墩
pinkeye /ˈpɪnkaɪ/ n. 传染性结膜炎
pinwheel /ˈpɪnwiːl/ n. 风车
pivotal /ˈpɪvətəl/ adj. 重要的，关键的
plight /plaɪt/ n. 困境
pneumonia /njuːˈməʊniə/ n. 肺炎
pole /pəʊl/ n. & v. 极点
pore /pɔː/ n. 孔隙

potable /ˈpəʊtəbəl/ adj. 适于饮用的
potassium /pəˈtæsiəm/ n. 钾
precipitation /prɪˌsɪpɪˈteɪʃən/ n. 降水
predator /ˈpredətə/ n. 捕食掠夺者
prioritize /praɪˈɒrɪtaɪz/ v. 按优先顺序处理
proliferation /prəˌlɪfəˈreɪʃən/ n. 增殖，扩散
prolong /prəˈlɒŋ/ v. 延长
propel /prəˈpel/ v. 推进
propeller /prəˈpelə/ n. 螺旋桨
proponent /prəˈpəʊnənt/ n. 支持者
proximity /prɒkˈsɪmɪti/ n. 接近

Q

quarry /ˈkwɒri/ v. 挖出

R

radial /ˈreɪdiəl/ adj. 放射状的
recharge /ˌriːˈtʃɑːdʒ/ v. 恢复
reclaim /rɪˈkleɪm/ v. 开垦，改造
relocate /ˌriːləʊˈkeɪt/ v. 重新安置
render /ˈrendə/ v. 导致
replenish /rɪˈplenɪʃ/ v. 补充，再装满
representative /ˌreprɪˈzentətɪv/ adj. 典型的
Reservoir /ˈrezəvwɑː/ n. 水库
responder /rɪˈspɒndə/ n. 响应者
riverbed /ˈrɪvəbed/ n. 河床
riverine /ˈrɪvəraɪn/ adj. 河的
rockfill /ˈrɒkfɪl/ adj. 填石的
rodent /ˈrəʊdənt/ n. 啮齿动物
rotate /rəʊˈteɪt/ v. （使）旋转
runoff /ˈrʌnɒf/ n. 径流

S

saline /ˈseɪlaɪn/ n. 含盐的，咸的
saturate /ˈsætʃəreɪt/ v. 使浸透，使充满，使饱和
sediment /ˈsedɪmənt/ n. 沉积物
sedimentation /ˌsedɪmenˈteɪʃən/ n. 沉积，沉降
seep /siːp/ v. 渗入，渗透
seepage /ˈsiːpɪdʒ/ n. 渗流
segment /ˈsegmənt/ n. 段
septic /ˈseptɪk/ adj. 腐败的
setting /ˈsetɪŋ/ n. 挡位
sewage /ˈsjuːɪdʒ/ n. 污水
shaft /ʃɑːft/ n. 竖井
shale /ʃeɪl/ n. 页岩，泥板岩

showerhead /ˈʃaʊəhed/ n.	莲蓬式喷头
sinkhole /ˈsɪŋkhəʊl/ n.	落水洞（尤指石灰石地面下陷，地表水消失于地下）
sleet /sliːt/ n.	雨夹雪
slope /sləʊp/ n.	斜坡，斜面
sludge /slʌdʒ/ n.	淤泥
solvent /ˈsɒlvənt/ n.	溶剂
sparsely /ˈspɑːsli/ adv.	稀疏地，贫乏地
spillway /ˈspɪlweɪ/ n.	溢洪道，泄洪道
spin /spɪn/ v.	（使）旋转
sprinkler /ˈsprɪŋklə/ n.	洒水装置，洒水车
staple /ˈsteɪpəl/ n.	主食
starve /stɑːv/ v.	挨饿
stewardship /ˈstjuːədʃɪp/ n.	管理工作
stock /stɒk/ n.	储备
stormwater /stɔːmˈwɔːtə/ n.	暴雨水
straddle /ˈstrædl/ v.	横跨
strangle /ˈstræŋgəl/ v.	扼死
streamflow /ˈstriːmfləʊ/ n.	流速及流水量
stretch /stretʃ/ v.	伸展，延伸
stymy /ˈstaɪmi/ v.	阻挠，妨碍
subdivide /ˌsʌbdɪˈvaɪd/ v.	细分
subirrigation /ˌsʌbɪrɪˈgeɪʃən/ n.	地下灌溉（通过地下多孔管道灌溉）
submain /ˌsʌbmeɪn/ n.	次干管（辅助干线）
submerge /səbˈmɜːdʒ/ v.	（使）浸没
submersion /səbˈmɜːʃən/ n.	浸没，淹没
subsidence /səbˈsaɪdəns/ n.	沉降，下沉
substrate /ˈsʌbstreɪt/ n.	基质
subterranean /ˌsʌbtəˈreɪniən/ adj.	地下的
subtropical /ˌsʌbˈtrɒpɪkəl/ adj.	亚热带的
suffocate /ˈsʌfəkeɪt/ v.	使缺氧
sulfur /ˈsʌlfə/ n.	硫
surmount /səˈmaʊnt/ v.	克服
surveillance /səˈveɪləns/ n.	监视，监督
swamp /swɒmp/ n.	沼泽
synchronize /ˈsɪŋkrənaɪz/ v.	使同步
synthesize /ˈsɪnθəsaɪz/ v.	合成，综合

T

tailing /ˈteɪlɪŋ/ n.	残渣
tailrace /ˈteɪlreɪs/ n.	尾水渠
tailwater /ˈteɪlˌwɔːtə/ n.	下游水（尾水；废水）
thrive /θraɪv/ v.	繁荣，兴旺
topographical /ˌtɒpəˈgræfɪkəl/ adj.	地形学的，地质的
topography /təˈpɒgrəfi/ n.	地势，地形
toxin /ˈtɒksɪn/ n.	毒素

transformer /trænsˈfɔːmə/ n.	变压器
transpiration /ˌtrænspɪˈreɪʃən/ n.	蒸腾作用
transpire /trænˈspaɪə/ v.	蒸腾
trapezoidal /ˌtræpɪˈzɔɪdəl/ adj.	梯形的
tributary /ˈtrɪbjʊtəri/ n.	支流
trickle /ˈtrɪkəl/ n.	细流
tropical /ˈtrɒpɪkəl/ adj.	热带的
trout /traʊt/ n.	鳟鱼
tsunami /tsuːˈnɑːmi/ n.	海啸
tunnel /ˈtʌnl/ n.	隧道，地道
turbidity /tɜːˈbɪdɪti/ n.	混浊度
turbulent /ˈtɜːbjʊlənt/ adj.	湍流的，汹涌的
typhoid /ˈtaɪfɔɪd/ n.	伤寒

U

underlie /ˌʌndəˈlaɪ/ v.	成为……的基础
underscore /ˌʌndəˈskɔː/ v.	强调
uniform /ˈjuːnɪfɔːm/ adj.	一致的
unsaturated /ʌnˈsætʃəreɪtɪd/ adj.	不饱和的
upstream /ˌʌpˈstriːm/ adj.	上游的
uranium /jʊˈreɪniəm/ n.	铀
utilize /ˈjuːtɪlaɪz/ v.	利用

V

vadose /ˈveɪdəʊs/ adj.	渗流的
valve /vælv/ n.	阀门
vapor /ˈveɪpə/ n.	水蒸气
variant /ˈveəriənt/ n.	变体
vegetation /ˌvedʒɪˈteɪʃən/ n.	植被
velocity /vɪˈlɒsɪti/ n.	速度
vertical /ˈvɜːtɪkəl/ adj.	直立的
visualize /ˈvɪʒuəlaɪz/ v.	设想
voluminous /vəˈluːmɪnəs/ adj.	大量的

W

waterborne /ˈwɔːtəbɔːn/ adj.	水传播的
waterlog /ˈwɔːtəlɒg/ v.	使（船等）进水，使浸透水
watershed /ˈwɔːtəʃed/ n.	分水岭
watertight /ˈwɔːtətaɪt/ adj.	水密的，不漏水的
weir /wɪə/ n.	堰堤，拦河坝
whereas /weərˈæz/ conj.	然而，鉴于
wicker /ˈwɪkə/ n.	枝条
windmill /ˈwɪndˌmɪl/ n.	风车

Z

zooplankton /ˈzuːəʊˌplæŋktən/ n.	浮游动物